ACID RAIN AND ACID WATERS

ELLIS HORWOOD SERIES IN
ENVIRONMENTAL MANAGEMENT, SCIENCE AND TECHNOLOGY
Series Editor: Dr YUSAF SAMIULLAH, ACER Environmental, Daresbury

ACID RAIN AND ACID WATERS

SECOND EDITION

Gwyneth Howells
Department of Zoology, University of Cambridge

Ellis Horwood

New York   London   Toronto   Sydney   Tokyo   Singapore

First published 1995 by
Ellis Horwood Limited
Campus 400, Maylands Avenue
Hemel Hempstead
Hertfordshire, HP2 7EZ
A division of
Simon & Schuster International Group

Printed and bound in Great Britain by
Bookcraft, Midsomer Norton

Library of Congress Cataloging-in-Publication Data

Available from the publisher

British Library Cataloguing in Publication Data

A catalogue record for this book is available
from the British Library

ISBN 0-13-336751-7

1 2 3 4 5   99 98 97 96 95

Table of contents

Preface

The reader may wonder why I should have rewritten a book published only four years ago, especially on a topic now so well covered by others, and one which might seem to have become less important now that many countries have adopted legal controls to limit emissions of acid-generating pollutants.

It seems, however, that scientific advances to document, quantify and understand acidification have even gathered pace, perhaps reaching the culmination of substantial programmes initiated in an earlier decade. Significant advances have been reported in recent conferences such as the 1990 international meeting in Glasgow, 'Acidic Deposition and its Nature and Impacts', and important national meetings sponsored in the United Kingdom by the Royal Society and in the USA by the American Chemical Society and by the consortium of government agencies which funded the National Acid Precipitation Assessment Program. These have provided a wealth of detailed information on mechanisms of ecological response to acidification, but most important, they have provided a necessary measure of the degree to which ecosystems have been affected by acidification and other factors. This has led to acquisition of reliable data for physical, chemical and biological systems from which sound conclusions can be drawn, and which provide a database for models that can anticipate future changes and their time-course with a variety of emission scenarios. Further work will be needed to validate or refute their predictions. Such an approach is necessary to judge the need and the 'value-for-money' of costly measures to control emissions, although this message seems often not heard by environmentalists and some politicians.

There is a further important aspect. The most recent UNCED convention in Brazil has expressed strong international concerns for maintaining the sustainability of our planet, while not yet formulating just how that might be achieved. The 'acid rain story' may be a model for the more sweeping aims of the Rio framework for global action (Agenda 21), from which politicians, administrators and scientists might take note, both of its confusions and failures and of the resolutions and successes of national and international acid rain programmes of the past 30 years. Perhaps it is naive to hope that the 'climate change story' will avoid the problems and pitfalls that were evident in following through the acid rain issue from initial disbelief and wild claims of disaster through to a more balanced and informed view.

ACKNOWLEDGEMENTS

I would like to acknowledge the helpful comments and suggestions of readers of the first edition, as well as the stimulating questions on acidification raised by final-year and postgraduate students who have attended my lectures on this topic. In addition, my knowledge of acidification processes and their reversal has continued to profit from my involvement in the Loch Fleet Project, a catchment liming project sponsored by the UK energy industry.

Permission for reproduction of various figures used in the text has been given by Chapman & Hall (Fig. 5.3), Royal Society of Edinburgh (Fig. 3.6, Fig. 5.7 and Fig. 5.12), John Wiley & Sons Inc. (Fig. 3.1), Warren Spring Laboratory (Fig. 2.3, Fig. 2.4, Fig. 2.6, Fig. 2.7) and CRC Press Inc. (Fig. 8.6). Other figures have been based on those from the sources attributed.

1

Introduction and scope

1.1 ACID RAIN—A *CAUSE CÉLÈBRE*

'Acid rain' has been a rallying cry of the 'environmentalist' and 'green' groups over
the past 35 years, and a powerful political, scientific and popular interest has
developed. The claim first put forward by a Swedish scientist, Svante Oden, in 1967
in a Stockholm newspaper was the basis of claims presented by Sweden (Royal
Ministry for Foreign Affairs and the Royal Ministry of Agriculture, 1971) to the first
United Nations Environment Conference in Stockholm in 1972. It declared evidence
for:

— a rapid increase in the acidity of rain in northern Europe;
— a parallel rapid increase in major rivers of Sweden;
— the growth of Swedish forests being in decline;
— numerous lakes and rivers having lost their fish populations.

This text will consider the extent to which these statements have been validated
by scientific investigation since 1972. In spite of more than three decades of intense
research activity, on both national and international levels, there is still confusion as
to how acidification is defined, to what degree it is man-made, and even what its
effects are and how they might be reversed or ameliorated. As a result, it is a topic
which might be said to have 'generated much heat, but not a great deal of light'.
Recent observations and assessments have improved understanding, underlining the
ecological nature of the acidification phenomenon—the pathway from source to
effects calls for analysis of atmospheric, terrestrial and aquatic environments. I hope
that this text will identify important questions that arise, and clarify some common
misconceptions by objective and scientific analysis. If it succeeds in stimulating its
readers to think independently and scientifically about a contentious and political
subject, it will have achieved a large measure of success.

Understanding acid rain is a serious challenge, since pathways through the
environment are complex (Fig. 1.1), and many disciplines are involved—some
knowledge of industrial processes, of atmospheric physics and chemistry, climate, soil

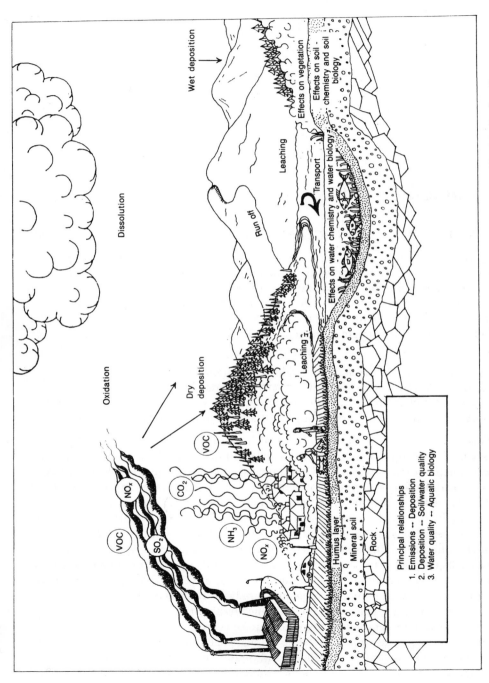

Fig. 1.1—Pathways from emissions through to targets of acidification and their principal relationships (partly after G. Persson (ed.), *Acidification Today and Tomorrow*, 1982). Acid-generating gases are sulphur and nitrogen gases (SO_2, NO_x, NH_3), carbon dioxide (CO_2) and volatile organic carbons (VOCs).

science, hydrology, land use and management, and terrestrial and aquatic ecology is called for. Alas, we are all specialists to some degree, and find it hard to apply the same standards of specialist knowledge and even probity to such a diversity of subject matter. We are often careful and critical of the area of science with which we are familiar, but sometimes ready to accept without the same critical standard what other specialists in other disciplines tell us of their conclusions. We may even assume naively that everything in print must be true! Our prejudices are often those declared by the pioneering predecessor of air pollution studies, Angus Smith (1872), who wrote 'a reason why I admire the results of...is one of a kind that affects all men and which may be excused. They agree with my own...'! Acknowledging such human frailties, in this multidisciplinary problem we have to achieve a consistent level of analysis, remembering that any false step in the logical development of the argument must invalidate further steps. Moreover, a poor level of accuracy in early steps of the chain of causality may well call in question the significance and even validity of later deductions.

Acid rain has sometimes been described as 'a cause looking for an effect'; although the acidity of rain and of some surface waters has long been recognized, the initial assumption of cause and effect relied heavily on making simple correlations between observations, often of a disparate nature. Such observations are often the beginning of scientific hypothesis formulation, but require rigorous testing by further observations and experiments to establish 'truth' by deductive and analytical reasoning. The application of a Popperian scientific approach to hypothesis rejection is a *sine qua non*. Correlations can never provide such proof, and even the most careful observations may need the support of credible mechanisms of action to be convincing. The concept of acidification and its consequent effects on a variety of ecosystems has thus developed from intuitive responses through an extensive series of observations of environmental phenomena, but it is also dependent on supportive and consistent findings from field or laboratory experimentation. The recent improvement in modelling these phenomena provides an opportunity to test the predictive hypotheses as well as to predict the benefits of actions to limit their cause. Unfortunately, where national and international pressures have forced the pace, conclusions have sometimes been reached in advance of sufficient observations and sound deductions, and there is bad as well as good science in this field.

1.2   ACID RAIN—A NEW PROBLEM?

The effects of 'acid rain' have been observed for centuries. A relationship between the vigour of plants and the health of the community with the development of early industry was first noted in the middle of the seventeenth century in England and France. Transnational exchange of pollutants between European countries was invoked even then (Smith, 1872; Cowling, 1980). As industrialization of Europe developed, other observers in England, Sweden, Austria and Germany confirmed the 'poisonous' nature of smoke emissions from various activities. Angus Smith, an English chemist working in Manchester in the 1850s as the first Inspector of Factories, was the first to analyse in a systematic way the chemistry of rain in the industrial heartland of Britain. He noted 'three kinds of air—that with carbonate of ammonia in the fields..., that with sulphate of ammonia in the suburbs..., and that with sulphuric acid in town air...'. Smith pursued this interest, publishing a book in 1872 entitled *Air and Rain: The Beginnings of Chemical Climatology*, and thus he can be considered the 'father' of acid rain—in fact he first used this term. It is worth noting that Smith believed that the air of industrial cities was contaminated by both industrial *and* domestic activities.

Independently of this industry-stimulated interest, another arose from promotion of agricultural productivity. Studies began in 1855 at the newly established Rothamsted Agricultural Station in Britain which showed that nutrient substances (specifically N compounds) in rain could contribute to crop needs for nitrogen. These interests stimulated a hundred years of rain sampling and analysis in Europe which still serves as a baseline from which later changes in rain composition can be deduced (Goulding et al., 1987).

Acidity was expressed as pH only at the turn of the century (Sorenson, 1909) and modern chemical concepts were only formulated scientifically around 1920 by Bronsted and Lewis who defined acids as proton donors and bases as proton acceptors (Skeffington, 1985; Stumm and Morgan, 1982). Advances in chemistry and analytical techniques followed, improving understanding of how rain and ambient air quality, as well as climatic variations, could lead to acid rain and in turn affect soils, forests and surface waters. There are good reasons for expecting that rain is acid to some degree due to washout of acid atmospheric constituents, but we have little direct evidence of its composition and acidity prior to industrialization. An Irish scientist, Conway (1942), published the first modern review of rain chemistry, and in 1948 the Swedish scientist Egner (Egner et al., 1955) initiated the first regional network of sampling, first in Sweden and later extended to some other European countries. At about the same time, Junge and Werby (1958) published data for monthly precipitation at various North American sites, although they did not measure pH.

Thus, evidence for an increasing trend in rain acidity since the 1950s depends on rather few data sets. The Swedish-initiated European Air Chemistry Network (EACN, later International Meteorology Institute network, IMI) was used by Granat and others (Granat, 1977, 1978) to show an increasing area of northern Europe subject to acid rain and sulphur deposition during the period from 1955 to 1970. Although Junge and Werby's North American data for 1955–56 did not include pH, this was estimated by Cogbill and Likens (1974) by ion balance and compared with more recent (1965–66 and 1972–73) data. Over the period 1955 to 1975, the EACN data for most stations showed no significant trend in acidity or sulphate, but about a quarter did so, of which only 10% showed a significant increase in both; about 50% showed a significant increase in nitrate concentrations. For the North American data, by comparing the later measured pH with the earlier calculated pH (with an analytical error of $\pm 10\%$), Cogbill and Likens concluded that an increase in maximum acidity and an enlarged area of acid deposition had occurred. Some concern has been expressed that other components of rain chemistry, notably ammonia and basic dusts, are not always reported but may have influenced the acidity, although not the sulphate records.

A more recent extension of the EACN network to a wider European scale by the Organization for Economic Cooperation for Europe (OECD) (Eliassen, 1978), and the later United Nations Economic Commission for Europe (UNECE) network of monitoring stations (European Monitoring and Evaluation Programme—EMEP, 1981), with a specific protocol for sampling and analysis, should provide better and more comparable records for the last decade. This activity, meeting the requirements of the 1979 UN Convention on Long Range Transport of Air Pollutants (LRTAP), should resolve any uncertainty about long-term trends in rain and its acid components (but see Chapter 3), as well as establishing the quantitative relationships of effects with acidifying emissions.

1.3 WHAT IS ACID RAIN?

We are all so familiar with 'rain'—at least in temperate regions—that it might seem unnecessary to define it. Yet it is a term without scientific precision or definition and this has led to some confusion. A more precise description is that it includes only *wet* deposition, distinguishing this from dry deposition of materials from atmosphere. But rain samples, as commonly collected, contain both wet and some dry materials deposited between rain showers. What is generally reported as 'rain' is better described as *bulk* deposition, although it may not represent the totality of atmospheric deposition. The dry component is notoriously difficult to sample, although it can be calculated, in theory, from atmospheric concentrations and their rate of removal; but even for commonly measured atmospheric components, this is difficult since deposition varies with surface characteristics and is influenced by diurnal, seasonal and climatic conditions.

'Rain' in common understanding is evidently the deposition of water vapour condensed from atmosphere, reaching the ground to complete the water cycle. However, it contains more than pure water of condensation, since as it falls from clouds through the atmosphere it will absorb gases and capture particulate aerosols during its passage. A major constituent of the atmosphere is, of course, carbon dioxide—a gas which is freely soluble in water to form carbonic acid—and thus the rain will be in equilibrium with carbon dioxide at the prevailing atmospheric pressure. This will reduce its pH (that is, increase its acidity) from a theoretical circumneutral value to one of slight acidity, or an acid concentration $[H^+]$ of about 1.5 μeq l^{-1}. Other natural constituents of the atmosphere will also dissolve in the falling rain, in particular the sulphate and nitrate derived from natural emissions of sulphur and nitrogen from a wide variety of sources—volcanic activity, electrical storms, bacterial denitrification, methanogenesis and sulphur reduction, volatile organic carbons and sulphur compounds emitted from vegetation, and algae (see Chapter 5). Other biologically produced gases such as ammonia (NH_3) may accept a proton to become ammonium (NH_4^+), reducing the acidity of rain. Within the terrestrial ecosystem it can be a nutrient, but also a source of acidity with twice the potential of nitric acid, adding to the acid loading of soils and drainage water.

In addition to gases, particulate materials (dusts and aerosols) can be entrained into the raindrop as it falls—such materials may increase (e.g. ammonium sulphate)

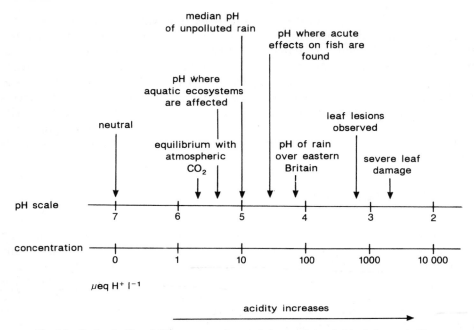

Fig. 1.2—Scale of pH and H$^+$ concentration, and the usual threshold of observed effects.

or decrease (e.g. calcium carbonate) the acidity of deposition. In maritime areas, sea salt aerosols are also dispersed to atmosphere and entrained in deposition; they are often ignored because they are considered neutral in respect of acidity, but they may have a significant role to play if such rain falls on areas of soils already deficient in bases (Chapters 7 and 8).

'Acid rain' is thus a term relating to the acids and potential acid generators in rain (and mists, fog and snow); it contains natural components as well as man-made ones and to some degree is almost always acid due to the presence of unneutralized anions from natural gases.

The acidity of solutions is expressed as 'pH' with a scale ranging from 1 (very strongly acidic) to 14 (very strongly basic) (Fig. 1.2). At pH = 7.0, the contribution of acidity is equalled by that of alkalinity. This scale is logarithmic, with each 'unit' denoting a tenfold change in acid concentration. Pure water (i.e. H$_2$O, lacking other constituents) exists only as a chemical concept and should have a pH of 7.0, but in the presence of carbon dioxide at atmospheric pressure and at a temperature of 20°C, such water will have a pH of 5.6. With the presence of other naturally generated gases (e.g. NO$_x$ and SO$_2$), and in the absence of neutralizing ions such as ammonium or limestone dusts, the natural rain pH is probably about 5.0, or a hydrogen ion concentration of about 10 μeq l^{-1}. This estimate is supported by measurements of rain in remote and unpolluted locations (Galloway *et al.*, 1982; Irving, 1991) (Fig. 1.3). In industrial regions, for example much of Europe and the North American continent, rain can be typically about ten times more acid than this, namely pH 4.0

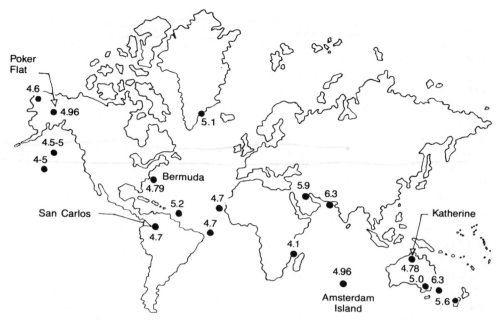

Fig. 1.3—Rain acidity (pH) found at remote sites throughout the world (after Galloway *et al.*, 1982; Irving, 1990). 'Pristine' rain at sites such as Amsterdam Island, in the Indian Ocean, has a pH ~ 5.0; some other sites unaffected by anthropogenic activity (South America) have rain pH < 5.0.

$[H^+] = 100 \ \mu eq \ l^{-1}$; European rain on average has a pH $\leqslant 4.2$. In rural regions, rain is more typically in the range of pH 4.5 to 4.9, although brief episodes of acidity are sometimes recorded, and pH values of < 3.5 have been recorded transiently at the initiation of a rain shower after a lengthy dry period. It should not be forgotten that rain is highly variable according to the antecedent and prevailing conditions.

The greater acidity of rain in urban and industrial areas arises because of the emission of sulphur and nitrogen gases from the combustion of fossil fuels, from a variety of fixed and mobile sources, or from other industrial processes like the smelting of ores. Other industrial emissions, including hydrocarbons, volatile organic carbons (VOCs), fluorine and chlorine, also have acidifying potential. The further oxidation of these emitted gases yields more soluble species which are taken up by rain as it forms as clouds and mist and falls. Atmospheric oxidation is not instantaneous (see Chapter 2) and deposition of acid gases peaks at a relatively short distance downwind of their source, while wet deposition of the soluble oxidation products is found at progressively greater distances and times, depending on climatic and atmospheric conditions. This oxidative process is not completely understood, since many minor constituents in the atmosphere, as well as the effects of humidity and sunlight, play a part (Chapter 2).

Emissions of natural and man-made acidic gases on a global basis are about equal, but in the more industrial northern hemisphere the latter account for more than 90%

of the total sulphur emissions. For nitrogen compounds, natural and man-made sources were thought to be about equal, but a recent increase in NO_x and NH_3 emissions due to human activity is likely to change this proportion. Sulphur, nitrogen and VOC emissions are about equal in the United States (Irving, 1991) but in Europe sulphur emissions are about the same as nitrogen gases and VOCs together. Chlorine, another acid gas, also reflects industrial activity, especially where high-chlorine coals are burnt or where plastic materials are manufactured or its waste disposed of. It should be acknowledged, however, that regional or global estimates of emissions are based on occasional samples and on speculation of a generalized rate of fuel consumption or overall industrial activity, and subject to considerable error. Recent work (e.g. Charlson et al., 1987) has suggested that natural biological sources have hitherto been underestimated, but recent estimates of natural sources of sulphur for the USA seem small, about 1–5% of man-made sulphur, while natural nitrogen sources are about 11% (Placet, 1990). The natural emissions attributed to tectonic activity are also stochastic and uncertain, but on a low resolution basis, estimates of emissions and deposition are roughly equivalent.

The primary concern for terrestrial ecosystems is the direct damage reported, or expected, due to direct exposure to noxious gases in air, such as SO_2, NO_x and ozone, rather than rainfall acidity. However, there is also a possibility that acid deposition, over time, has depleted the supply of essential trace metals required for plant health from poorly endowed soils. In contrast, effects on soils and aquatic ecosystems are related to deposited or derived acidity, not to the concentration of sulphate or acidity of rainfall. This book will focus on effects in surface waters, as mediated by transfer through vegetative canopies and soils, although recent concern for 'forest dieback' in Europe and elsewhere has prompted a review of its causes.

In addition to effects on the natural environment, acid rain is judged a cause of potential detriment in human health, of reduction in atmospheric visibility and of structural damage. These aspects have been reviewed authoritatively for conditions in the northeastern United States by the agencies forming the National Acid Precipitation Assessment Program (NAPAP), to which the interested reader is directed (Irving, 1991). Other reviews of these aspects are to be found in UNECE reports (UNECE, 1982a, b, 1984, 1985, 1986) and for United Kingdom conditions in Mellanby (1988).

1.4   ACID RAIN OR AIR POLLUTION?

It has been less than helpful to scientific analysis and understanding that the term 'acid rain' has been popularized to include all gaseous air pollutants, including ozone (e.g. Rossi, 1984). It is also assumed that 'acid rain kills forests and fish' although forests react more to atmospheric gases and fish do not live in acid rain. The contribution to acid rain of natural sources of acidifying gases in the atmosphere, and of natural phenomena of acidification, is seldom acknowledged, although critical to consideration of the expected benefit from emission control.

While the acidic gases may dry-deposit to add to the total input of acidic inputs to ecosystems, they do not necessarily influence the acidity of rain, being of low solubility in their partially oxidized state. Their effects on the above-ground terrestrial environment are related to their chemical nature, not their potential for acidity; they cause direct effects if present in concentrations toxic to vegetation or damaging to man-made structures. These effects are not the same as those of acidity, direct or indirect, on vegetation, soils and aquatic ecosystems. While the sources of some gaseous pollutants and of acidic materials in rain may be the same, their fates, effects and remedies may be different. Some secondary air pollutants (e.g. ozone) may also have significant damaging effects on vegetation yet do not affect the acidity of rain directly. In particular, the role of ozone in environmental acidification has yet to be satisfactorily resolved (Derwent *et al.*, 1987). Paradoxically, while ozone depletion may bring other hazards, its reduction may reduce atmospheric oxidation of sulphur but enhance NO_x levels.

1.5   THE SEQUENCE OF EVENTS—SOURCE TO TARGET RESPONSE

In considering the formation and fate of acid rain as an integrated phenomenon, we must consider the sequence from source to target, developing wherever possible, quantitative relationships:

— between source terms and deposition at distant sites;
— on the fate and dynamics of transfer of the acidic components;
— on the dose–response of target species or communities;
— on the overall consequences in aquatic ecosystems;
— on the timescales over which effects are manifest;
— on the potential for recovery, and its timescale.

The following chapters will follow this course.

2

Emissions, dispersion, deposition

2.1 EMISSIONS—SOURCES, POLLUTANTS, FORMS

Pollutant emissions leading to acid rain are predominantly the oxides of sulphur and nitrogen. **Sulphur dioxide** (SO_2) is formed by the oxidation of sulphur present in fossil fuels, the sulphur initially formed by biological activity, as well as some from geological materials. Smelting of ores also produces sulphur oxides, while the old 'Kraft' process for paper manufacturing is another source. Natural events such as volcanic and tectonic activity may also be an important, although discontinuous, source; sulphate in rain collected downwind of a volcano in Hawaii is high (and matched by acidity increase) following activity but often undetectable at other times (Siegel *et al.*, 1990). In contrast, intermittent volcanic eruptions at Mount St Helens in 1980 are reported to have had little effect on acid components (Lewis and Grant, 1981) although chemical and biological changes in lakes of the area were substantial (Wissmar *et al.*, 1982a, b). Together with emissions from biological processes, such as the production of dimethyl sulphide and hydrogen sulphide, these contribute 1–5% to atmospheric emissions in the United States; values are higher in the summer and near the coast (Irving, 1991). In the northern hemisphere man-made sources clearly dominate emissions, while in the tropics or the southern hemisphere natural sources are dominant. A significant emission occurs of natural biogenic sulphur compounds, such as H_2S, from decaying swamp vegetation, or of dimethylsulphide (DMS) from the activity of marine phytoplankton or macroalgae (Charlson *et al.*, 1987), estimated at 3 mmol m^{-2} yr^{-1}.

Nitrogen oxides (NO and NO_2, usually expressed as 'NO_x') are generated during the process of combustion by oxidation of nitrogen present in air. The degree to which this occurs is dependent on combustion temperature; some N gases are also generated from the chemical industry. The nitrogen content of the combusted materials (e.g. organic wastes) also contributes to the NO_x emission. Natural phenomena, such as N fixation during electrical storms, or biological N fixation followed by denitrification, contribute significant quantities. In North America, these natural sources are about 11% of the total N emissions; they are greater in summer and may be a significant influence in areas with low man-made emissions (Irving, 1991). Another N compound, **ammonia** (NH_3), is also important; it is alkaline in rain but with its subsequent conversion to ammonium (NH_4^+) and nitrate (NO_3^-) in water, it adds acidity to soils and surface waters. Recent estimates of ammonia emissions in Europe indicate that about 80% comes from livestock waste (Fowler, 1992) and it has been shown to be a substantial acidifying agent in the Netherlands (Fowler, 1992). Better resolution of the relative proportions of natural and man-made emissions of N and their potential for acidification of soils and waters is needed.

Emissions of **non-methane hydrocarbons** (NMHC) from industry, from leakage of natural gas and from vegetation are also now considered significant contributors to **volatile organic carbon** (VOC) in the North American atmosphere, and to a lesser extent in Europe—again natural emissions are greater during summer. It is believed that this influences the seasonal production of oxidants in the atmosphere, although the emission estimates are imprecise (Irving, 1991; Irwin et al., 1990).

A further acidic emission associated with some fossil fuels and other industrial activities, including waste disposal, is **chlorine** (Cl) gas. About 75% of chlorine emissions in Europe are from coal combustion (Irwin et al., 1990). Chlorine is readily dissolved in rain or cloud vapour and can influence rainfall acidity close to its sources. A detailed emissions inventory is not yet available, although in the United Kingdom the annual emission from coal is as much as 0.3 Mtonnes (Irwin et al., 1990). Much of the chloride in rain is, of course, derived from suspended sea salt, especially in maritime regions. All rain in western Europe is affected by marine salts and marine ions dominate rain chemistry in western UK sites. In North America, sea salts are unimportant at sites > 200 km from the coast (Irving, 1991). Since sea water is slightly alkaline (pH = 8.2) it is generally assumed that sea salt aerosols do not influence rain acidity although their uptake and redissolution in rain may do so; they certainly contribute a significant fraction of sulphur deposition in maritime regions and influence the chemistry of runoff (see Chapters 6 and 7).

In addition, the increasing atmospheric burden of **carbon dioxide** (CO_2) will also influence the carbonic acid present in rain, governed by the ambient temperature and pressure; a doubling of the present P_{CO_2} (partial pressure) in atmosphere would approximately double the acidity of rain due to carbonic acid (Stumm and Morgan, 1981).

When acidifying gases are generated they are dispersed and diluted downwind of the source; chemical oxidation in the atmosphere proceeds during this process (Fig. 2.1). These reactions convert the oxides of sulphur and nitrogen to sulphuric and nitric acid; the oxidizing agents are varied. In dry air, SO_2 gas is converted to SO_4 by OH radicals, while in moist air, the sulphate forms cloud droplets which may be washed out by rain. Similarly NO_x reactions differ between night and day. The atmospheric oxidation process (see the box and text below) is complex, limited by the oxidants present, and is influenced by light and moisture; it is often slow, so that removal of the atmospheric burden by rainfall may be at some considerable distance from the emission sources.

For the most part, emissions are contained within the 'mixed layer' of atmosphere, the envelope of well-mixed air close to the Earth's surface, typically about 1 km deep.

PRINCIPAL CHEMICAL REACTIONS OF SULPHUR AND NITROGEN GASES EMITTED TO ATMOSPHERE

Generation of oxidants:

NO is the major nitrogen oxide from combustion:

$$NO + O_3 \rightarrow NO_2 + O_2$$

In sunlight, where the rate depends on the solar intensity:

$$NO_2 + h\nu \rightarrow NO + O$$

$$O_2 + O \rightarrow O_3$$

In the presence of hydrocarbons:

$$RH + 2NO + 2O_2 \rightarrow carbonyl + 2NO_2 + H_2O$$

Photolysis of ozone in presence of water vapour produces hydroxy radicals (OH), and of carbonyl in presence of NO produces HO_2; recombination of hydroperoxyl radicals produces hydrogen peroxide:

$$HO_2 + HO_2 \rightarrow H_2O_2 + O_2$$

Dry oxidation of S and N oxides:

Sulphur dioxide reacts with OH radicals:

$$SO_2 + OH \rightarrow HSO_3$$

$$HSO_3 + O_2 \rightarrow HO_2 + SO_3$$

Nitrogen dioxide reacts with OH:

$$NO_2 + OH \rightarrow HNO_3$$

(this is ten times faster than the S reaction).

Wet oxidation of sulphur:

SO_2 dissolves in rain as bisulphite and H^+:

$$SO_2.H_2O \rightarrow HSO_3 + H^+$$

Oxidation proceeds to sulphate and more H^+:

$$HSO_3 \rightarrow SO_4 + H^+$$

This is slow, but is catalysed by oxidants and transition metals.

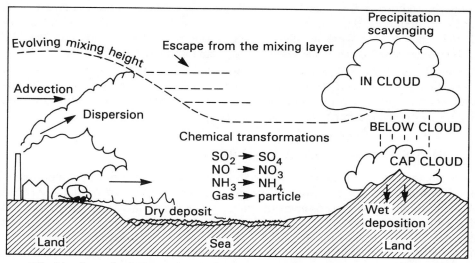

Fig. 2.1—Generalized scheme of reactions and processes in the troposphere and stratosphere
(after Fowler, 1992).

Mixing varies diurnally so conditions may not be uniform, making it difficult to balance emissions with deposition within a reasonable area. Some emissions may possibly escape to the troposphere and some material may possibly be anomalously distributed, as seen in the dispersion of the Chernobyl radioactivity.

2.2 EMISSION DISPERSAL

Emissions from point sources such as chimneys, or from multiple sources within industrial or urban areas, can be dispersed downwind for hundreds of kilometres within the mixed polluted layer of atmosphere. The same forces disperse natural materials such as sea salts and the dusts and gases generated by dust storms and volcanic activity. Thus, materials from both natural and man-made sources can transgress national boundaries before being deposited. During this dispersion, the constituents are progressively diluted, typically by a factor of 10 000 within 10 to 20 km from a source of height 220 m. Oxidation proceeds during this phase at rates estimated as only 1–3% per hour, influenced by atmospheric conditions of sunlight, humidity and the presence of catalysing materials, while deposition is influenced by the Earth's surface characteristics. In dry weather in the UK, nearly half of SO_2 emissions remain airborne for more than 24 hours, during which time they are likely to have travelled about 600 km (at wind speed 7 m s^{-1}, Goldsmith *et al.*, 1984).

It is clear that estimates of atmospheric removal will be highly variable, according to conditions prevailing along the path of a plume, and to diurnal, day-to-day and seasonal conditions which govern the degree to which emitted materials are dispersed, oxidized and deposited. But the height of the emission sources, contrary to popular conception, has relatively little consequence on deposition—it might result in a greater initial deposition from a smaller plume at lower height and so a larger near-field deposit, but as soon as the plume is entrained into the boundary layer it behaves in the same way as from a taller chimney. At long distances, say 1000 km, the difference is estimated to be less than 10% (Crane and Cocks, 1987).

Thus, in the near-field, dry deposition dominates, but with progressive dispersal and oxidation downwind more soluble constituents are formed so that in the far-field wet deposition becomes more important. At any site, however, both local and distant sources will contribute to deposition.

2.3   DEPOSITION AND REMOVAL FROM ATMOSPHERE

Two principal mechanisms are responsible for removal of the atmospheric burden—**dry deposition** and **wet deposition**. Materials present as gases and partially oxidized aerosols are deposited dry, whereas the fully oxidized materials are more soluble and deposit during rain episodes. Some aerosols may be entrained in rain as it falls, and dry particulates deposit on to surfaces of rain samplers between rain events, and influence the composition and acidity of such 'bulk' rain samples. Thus both dry and wet deposition are important routes for atmospheric fallout but their relative proportions depend on the time and distance from source, as well as seasonal atmospheric conditions. Wet deposition is greatest in areas of high rainfall and may be > 70% in orographic areas (Irwin *et al.*, 1990), while in North America dry and wet deposition are about equal over the year at many sites (Irving, 1991).

When 'rain' is collected in a standard (bucket-type) gauge it contains the concurrent wet component of deposition but also dry deposition falling between rain events, or entrained by rain as it falls through the atmosphere—a variable component. Problems with excluding all dry deposit from rain samplers have proved so difficult that there are few good 'wet-only' records published. In almost all published data the 'rain' refers to **'bulk'** deposition—not necessarily the total of dry + wet deposition, nor a consistent fraction, although considered a practical approximation. Dry deposition is also difficult to measure directly and it is almost always estimated on the basis of measured atmospheric gas concentrations and an estimated **'deposition velocity'** of transfer to surfaces, varying with their texture and nature. The impaction of mist, fog and cloud—termed **'occult'** deposition—on to surfaces such as tree canopies and at high altitudes in temperate regions is also under-represented in bulk rain samples (Roberts *et al.*, 1983; Fowler, 1992). On hill tops, rain concentrations are increased, explained as scavenging of droplets in frontal clouds (the 'seeder–feeder mechanism', Fig. 2.2). In wet canopies the take-up of SO_2 is enhanced by the presence of ambient NH_3, promoting oxidation to SO_4^{2-} and its neutralization (Fowler, 1992)—a process of co-deposition; this has been shown to be significant in the Netherlands. Both climatic conditions and the roughness of the canopy determine the effectiveness with which the atmospheric materials are captured.

In addition, the different deposition mechanisms ('rainout', 'washout' and scavenging) have a different effectiveness of pollutant capture. This is evident in samples of wet deposition below tree canopies, not represented by open 'bulk' collectors set out in the open. The significance of episodic conditions is also obscured by methods

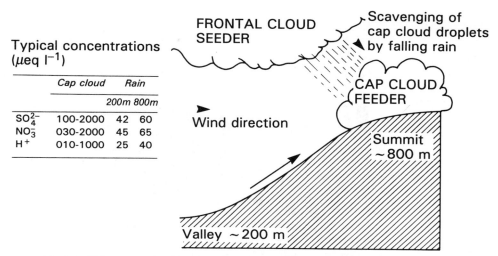

Typical concentrations
(μeq l^{-1})

	Cap cloud	Rain	
		200m	800m
SO_4^{2-}	100-2000	42	60
NO_3^-	030-2000	45	65
H^+	010-1000	25	40

FRONTAL CLOUD SEEDER

Scavenging of cap cloud droplets by falling rain

CAP CLOUD FEEDER

Wind direction

Summit ~800 m

Valley ~200 m

Fig. 2.2—Enhancement by the 'seeder–feeder' mechanism of deposition and concentrations at higher altitude (after Fowler, 1992).

based on accumulated samples over time, although practical as a routine.

Thus, although it might seem a simple matter to judge the amount and effects of acid deposition at a site by measuring the quantity of rain and its chemistry, this provides only an approximate estimate of the true value. In the following text 'rain' refers to the customary bulk samples as reported and will underestimate the total.

It is also important to distinguish between deposition **concentrations** and **fluxes**, the latter being the product of volume and concentration. In detailed studies of national monitoring network data in the UK (Irwin *et al.*, 1990) a different picture emerges of regional differences in rain composition (concentrations) and of deposition (fluxes) between eastern and western areas. The area of maximum rain acidity ($>80\ \mu$eq l^{-1}, pH <4.1) lies in the Midlands, close to major sulphur emission sources, whereas wet deposition of acidity is highest in western and northern Britain, mostly distant from industry (Fig. 2.3), a pattern also seen for the acid components of rain (sulphate, nitrate and chloride). This distinction is also important in estimating the source of acidifying components—for (non-marine) sulphate, for example, the highest concentrations in rain are from easterly or northeasterly quarters and are associated with a small rainfall volume, while wet sulphate deposition is predominantly from the southwest and is associated with heavy rain events (Fig. 2.4) (Irwin *et al.*, 1990). This westerly deposition represents air masses which have not passed over major source areas upwind.

The removal of atmospheric sulphur and nitrogen gases in wet conditions is more effective than in dry conditions. The S content of rain far exceeds that which can be accounted for by the solution of the gas, sulphur dioxide, in water; in fact its dissolution is self-limited by the increasing acidity. Atmospheric oxidation of the sulphur dioxide gas to sulphate is necessary to explain the higher concentrations

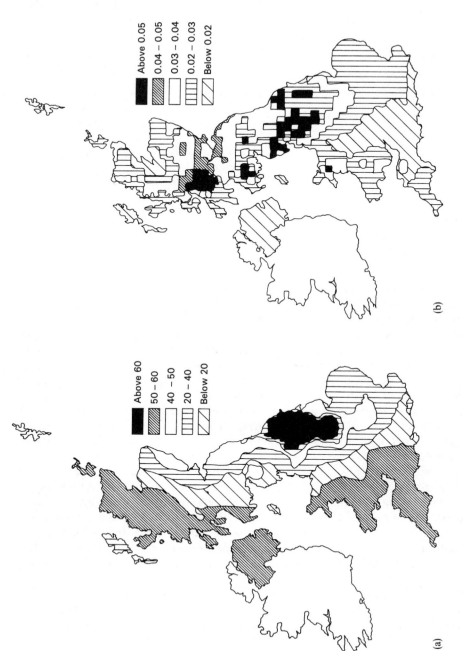

Fig. 2.3—Regional pattern of (a) acid rain concentrations (μeq H$^+$ l^{-1}) and (b) annual wet deposition (g H$^+$ m^{-2}) across the UK (after Irwin *et al.*, 1990).

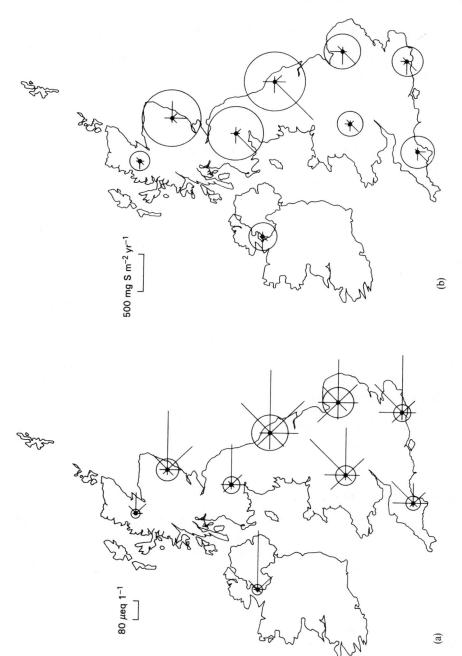

500 mg S m^{-2} yr^{-1}

(b)

80 µeq l^{-1}

(a)

Fig. 2.4— Sectoral non-marine sulphate contributions to rain concentrations (µeq SO$_4$$^{2-}$ l^{-1}) and to annual wet deposition (g S m^{-2}) (after Irwin et al., 1990).

observed—first to bisulphite (SO_3^-) and then to more soluble sulphate (SO_4^{2-}). Similar oxidation reactions for nitrogen species are probable, but are not so well documented. Major oxidizing agents are hydrogen peroxide and ozone, catalysed by manganese and iron, while atmospheric oxygen has only a minor role. While hydrogen peroxide acts quickly, its concentration is usually low, limiting its effect; ozone reacts more slowly because it is scarcely soluble and is adversely related to acidity. In remote areas, oxidation is the major mechanism for removal of the SO_2 gas from the atmosphere, but in more polluted areas, such as most of Europe and North America, gaseous SO_2 concentrations exceed the capacity of the oxidants available, so it is effectively controlled by regeneration of hydrogen peroxide (Crane and Cocks, 1987) (see box earlier in this chapter).

Another route of oxidation is via reaction with hydroxyl (OH) radicals to sulphur trioxide (SO_3^-) and thence to sulphuric acid (H_2SO_4). This seems less important as the rate of oxidation is too slow to account for the measured levels of sulphate in rain. An important issue is whether the concentration of sulphate in rain is proportional to the concentration of sulphur dioxide in the atmosphere (and to its emission); although there is some proportionality, the presence of other atmospheric constituents, as well as prevailing conditions and the infrequency of rain events, which all have a part to play, makes a simple relationship unlikely. Indeed, the wet deposition of sulphur is not linearly correlated with emissions in areas of high SO_2 because of the restricted availability of oxidants (Irving, 1991). These findings have important implications with regard to expected changes in bulk deposition of sulphate following changes in sulphur emissions.

The primary nitrogen emission, nitric oxide (NO), is not significantly dry-deposited; it is oxidized in the atmosphere to nitrogen dioxide (NO_2), nitrate (NO_3^-) or nitric acid (HNO_3). Since nitrate is freely soluble this is the major pathway for its removal from the atmosphere. The nitrogen oxides are also involved in oxidant production in the atmosphere (including formation of ozone), so again a proportional relationship between emissions and wet deposition is unlikely. In addition the role of N fixation and denitrification by bacteria and algae is not well established on a regional field basis, and poor estimates of ammonia generation and oxidation further confuse the relationship of nitrogen emissions and deposition (Irwin et al., 1990).

2.4 MODELLING ATMOSPHERIC PROCESSES

The slow rate of atmospheric oxidation reactions suggests that several days within a dry plume are necessary to account for the measured amounts of sulphate and nitrate in rain. This makes it difficult to track the origins and pathways of polluted air 'parcels' en route by sampling and measuring air chemistry and of rain in target areas (Clark et al., 1986). On the basis, however, of information on air concentrations and transformations, coupled with understanding of the behaviour of air masses, credible models relating emissions to deposition have been formulated (e.g. Fisher, 1983). These have been developed for Europe internationally, through UNECE/EMEP and OECD, and between the USA and Canada, over more than two decades. Data collection has been standardized and coordinated through EMEP, with participation from national monitoring networks in North America, the UK and other European countries. The data on deposition, along with those for emissions and meteorological data, have been used to predict deposition over North America (Regional Acid Deposition Model, RADM) and Europe (Eliassen and Saltbones, 1983; Chang, 1990; Dennis, 1990). Analysis of EMEP data has shown that zonal atmospheric pressure differences can provide an explanation of differences and trends in rainwater quality, with a timescale of about three decades (Davies *et al.*, 1992). The principal 'depression track' running through the North Sea to the Baltic provides transport from Europe through to Scandinavia.

The recent establishment and operation of sampling networks has greatly improved the quantity and quality of the data available, but it is widely admitted that the data needs of most models are larger than these networks can provide. A variety of statistical and process-type models has been explored, but none seems ideal (Irwin *et al.*, 1990). Models are designed to meet many needs such as estimating concentration or deposition patterns, predicting source–receptor relationships, modelled and observed processes, predicting the consequences of alternative emission scenarios, and investigating specific events to predict worst-case conditions. They almost all include meteorological processes such as advection, mixing depth and inputs and losses from the mixing layer, chemical transformations in the atmosphere, wet and dry deposition, and the contribution from background. Comparison between models and their validation by observations has shown that empirical (statistical) models and trajectory (Lagrangian) sulphur transport models can be used for reasonable predictions of long-term average concentrations of regional deposition. However, the prediction of long-range nitrogen deposition is handicapped by the limited measurements available (Irwin *et al.*, 1990). No models are yet suitable for short-term or episodic events.

Rather simple models (accuracy $\pm$ 50%) have been used to relate national emissions to deposition without much recognition of their non-linear relationships; they give a false impression of certainty. They have led to the formulation of 'blame' tables for the export/import of sulphur, firing much of the international arguments for emission control (Highton and Chadwick, 1982). However, it has become clear that wet sulphate deposition in Europe exceeds the predictions of the EMEP (trajectory) model, i.e. it cannot all be accounted for by declared European emissions. The excess, termed 'background' deposition, is 15–60% of the total (Fig. 2.5), possibly reflecting poor estimates of natural and industrial emissions, perhaps longer-range transport or complex recirculation of European emissions, or some exchange of atmospheric constituents between troposphere and stratosphere. This 'background' is highest in western coastal areas of the UK and Scandinavia subject to orographic deposition (Crane and Cocks, 1987; Irwin *et al.*, 1990), and is also recognized in the USA (Irving, 1991).

Reassessment has led to some reservations about the export/import approach. While earlier estimates for the Scandinavian countries suggested that the UK contributed substantial quantities of sulphur to them, recent EMEP estimates suggest that this could be only about 8% of the total deposited in Norway and about 3% in Sweden (and <1% in western Germany). About 40% of European emissions are

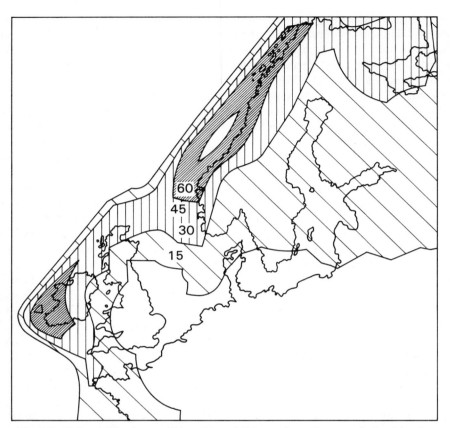

Fig. 2.5—Estimated 'background' deposition of S in northern Europe, calculated by the
EMEP model (after Crane and Cocks, 1987). Isopleth values refer to the percentage in excess
of the calculated deposition.

also exported out of the region (Iversen *et al.*, 1989). In North America there is a
significant export (10–30% of regional emissions) of sulphur from the eastern USA
to eastern Canada, and a reciprocal but smaller export (about half as much) from
eastern Canada to the USA. Indeed, there is a significant eastwards advection
(15–45%) across the Atlantic (Galloway and Rodhe, 1991). The uncertainty of
these estimates is thought to be about 20%. However, the eastward flux of sulphur
to Europe is thought to be less than a tenth of the 3–4 Mtonnes yr^{-1} leaving
North America.

Similar intra-national fluxes have been attempted for N species, although its more
complex chemistry in the atmosphere and uncertain estimates of source strengths
make these even less precise. In North America, possibly 30% is transferred from
the northeastern USA to Canada and 25–40% goes eastwards across the Atlantic
(Iverson *et al.*, 1989; Galloway and Rodhe, 1991), perhaps about 1 Mtonne N yr^{-1}.
Both reduced and oxidized nitrogen forms from European emissions may contribute

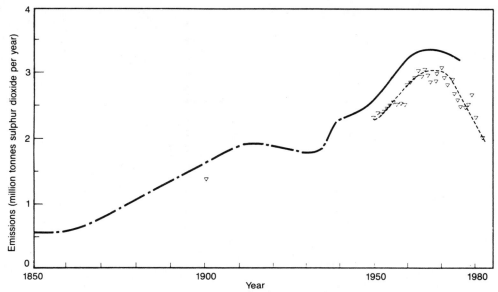

Fig. 2.6—Estimated trends in UK fuel consumption through 140 years, expressed as S
emissions (after Chester, 1984 ——; Irwin *et al.*, 1990 △△; Fisher, pers. comm.).

to the productivity of the North Sea; the UK contribution is about a third of this.
However, estimates of NH_3 budgets, emitted mostly from Denmark, the Netherlands
and France, suggest that this may be greater than the NO_x export (Soderlund, 1977).

It should be remembered that uncertainties in the model lead to substantial
inaccuracies (by a factor of two for S budgets) which can only be resolved by better
field data and better-validated models. In particular the 'sources' and 'sinks' of
sulphur and nitrogen in oceans are not well documented, and only a few estimates
have been made on an ocean or regional basis. Some ethical questions also arise—
should the 'export' be considered politically on a GNP or population size rather
than a national one; should 'imports' be considered on a national area basis or in
relation to conditions such as regional/local sensitivity?

2.5   TRENDS IN EMISSIONS AND DEPOSITION

Emissions in the UK and some other European countries have been estimated on
the basis of fuel consumption over the last 140 years (Fig. 2.6). In Britain these
estimated emissions rose from about 0.5 Mtonne of sulphur per year in the mid-
nineteenth century, levelling out between 1910 and 1920, and then rising to a peak
of >3.0 Mtonnes in 1965. Since that time UK emissions have dropped steadily; they
were <2.0 Mtonnes in 1985, similar to the 1920s level (Irwin *et al.*, 1990), and about
1 Mtonne in 1990 (Fisher, pers. comm.), a fall of about 60% since the peak. The
decline is attributed to the changing style of industry, energy conservation, and the
use of 'clean' fuels such as gas and nuclear energy.

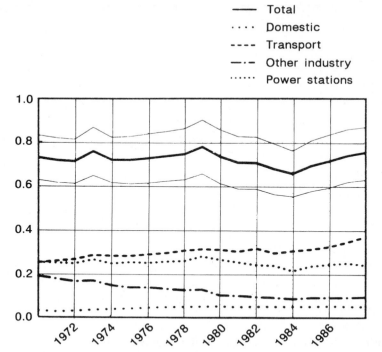

Fig. 2.7—Trends in estimated annual NO_x emissions (Mtonnes N yr^{-1}) in the UK (after Irwin *et al.*, 1990). The upper and lower bands around the 'total' curve represent 95% confidence limits of the data.

In Britain, the pioneering Clean Air Act of 1957 brought great benefits, especially in urban and industrial areas, by limiting not so much the levels of sulphur emissions, but that of 'smoke' which had been implicated along with sulphate aerosol in the 1952 London smog disaster. The Act brought about a sharp decline in the SO_2 in urban air, largely by improving the technology of domestic heating and greater efficiency of power industry generation as well as the greater dispersion and dilution of emissions from tall stacks. At the same time, new sources of energy became available with the exploitation of oil and natural gas in the North Sea. The old primary industries of ore smelting and iron and steel production have declined, and rail transport is now almost entirely powered by diesel fuel and electricity rather than the direct combustion of coal (Mellanby, 1988). These changes have been matched in other European countries in later years, although in eastern Europe, high-sulphur coal, heavy industry and low thermal efficiency still lead to a high level of sulphur emissions there.

The picture for nitrogen emissions is less optimistic, however. Total UK emissions of NO_x rose from 0.3 Mtonne per year in 1945 to 0.6 Mtonne per year in 1980, largely as the result of increased motor traffic (Derwent *et al.*, 1989; Fig. 2.7). Ammonia emission estimates are less certain, with increases in Europe judged to be about 50%

between 1950 and 1980 (Skeffington and Wilson, 1988). This trend continues in the UK, unabated by effective control since a reduction in industrial sources has been offset by the increase in traffic and more intensive agriculture. This is reflected in a significant increase in the nitrate and ammonia content of rain (Brimblecombe and Stedman, 1982; Skeffington and Wilson, 1988).

The trends in emissions in Europe suggest that sulphur emissions rose steadily throughout this century but doubled from the 1950s to 1965–70 and then declined, while nitrogen emissions have continued to increase. These changes are only broadly reflected in the composition of rain (see Chapter 3) with lower sulphate and higher nitrate concentrations (Barrett *et al.*, 1983) but with little change in acidity. At some UK sites remote from industrial sources and receiving a larger fraction of 'background' source material, sulphate deposition and the acidity of rain have declined (Fowler and Brimblecombe, 1988; Battarbee *et al.*, 1988), but the correlation is weak and the same change is not evident for other UK sites (Barrett *et al.*, 1987; Irwin *et al.*, 1990). In Scandinavia, an observed decline in sulphate in deposition and in surface waters is also not matched by rising pH (Henriksen *et al.*, 1988; Seip *et al.*, 1990); 30 years of rain sampling in Sweden suggest that changes in sulphate, acidity and alkalinity have been statistically close to zero (Sanden *et al.*, 1987), even though emissions of SO_2 are less than half of the 1970s level. There is stronger evidence that the increasing N emissions over the past two decades have resulted in higher nitrate concentrations in rain and surface waters.

In the USA all three important precursors of acid rain, SO_2, NO_x and VOCs, increased from 1900 to 1970, when the first Clean Air Act was enacted. Sulphur dioxide increased by a factor of three (9 to 28 Mtonnes), NO_x tenfold (2 to 18 Mtonnes), and VOCs twofold (Irving, 1991). Over the two decades since emission controls began to operate, emissions of all three pollutants have been reduced, by about 28% for SO_2, 9–15% for NO_x, and 27–39% for VOCs. US rain shows recent decreases in both nitrate and sulphate, but no change in acidity, perhaps related to a fall in base cations (Irving, 1991). The US observations confirm the complex and uncertain relationship between lower acid emissions and rain chemistry. Interpretations are by no means simple, since year-to-year changes in climate (and the amount of rainfall) can be important, as well as the variable amounts of neutralizing calcium and ammonium affecting the balance of anions and cations in rain.

Notwithstanding these scientific problems, the recently revised US Clean Air Act (1990) will, *inter alia*, reduce SO_2 emissions from power plants by 50% by the year 2000 and total annual emissions will be capped at 8.9 Mtonnes (<50% of 1980 levels). These plants will be given tradeable 'allowances' based on fixed emission limits. NO_x reductions will also be achieved, in this case by improved performance standards (Helme and Neme, 1990; Claussen, 1990).

2.6   UNRESOLVED ISSUES

There is general expectation that reductions in sulphur and nitrogen emissions will result in reduced acid deposition. While there is some evidence that declining sulphur emissions are matched by rising nitrogen emissions, unchanging rain acidity is not reassuring and simple responses following emission changes may not follow, because of oxidant limitations. Further, the timescale of any change seems to be longer than that expected from atmospheric phenomena.

Some further questions arise

What is the true level of 'background' deposition and what are its origins?
What is the quantitative effect of ammonia emissions or of nitrogen released by denitrification?
How universal are the few observations of greater acid deposition at higher altitude or where occult deposition seems to predominate? What is their quantitative significance?
What are the crucial chemical processes involved in ozone formation or removal? Will ozone or other oxidants, or increasing NO_x and VOCs, limit the benefits of lower emissions?
What is the significant role of other atmospheric constituents such as chlorine, hydrocarbons or other gases?

So right at the beginning of our chain of consequences from source to target we are faced with uncertainties about the origins and formation of acid deposition. While scientific research and extensive monitoring over the past two decades has done much to improve techniques of sampling and measurement, there is still much to be done in developing better understanding so as to clarify, by observations, experiment and modelling, the crucial but elusive relationships between emission sources and deposition which must be the cornerstone of an effective control policy.

2.7 SUMMARY AND CONCLUSIONS

Over recent years much new information has been generated about the behaviour of emitted gases, their atmospheric transformations and interactions, and about mechanisms of the removal of gases and aerosols from the atmosphere. National and international cooperative programmes have done much to provide data of a consistent and comparable character so that models of regional dispersion and deposition have been developed, and sometimes validated.

In summary

Emissions of sulphur and nitrogen are the major primary precursors of acid deposition. The greater part of these emissions in the northern hemisphere is man-made.
Secondary components of air pollution include ozone and other photochemical oxidants in the atmosphere, responsible for the progressive oxidation of S and N species during the downwind dispersion of emissions.
VOC emissions are of increasing concern since they influence atmospheric reactions. Increases in atmospheric CO_2 levels may also be important.
Atmospheric oxidation reactions are slow, and are both self-limited and acid-limited; oxidation increases solubility and the likelihood that its products will be

removed in rain as much as 1000 km distant from sources, dependent on wind strength and trajectory.

Dry and wet deposition are the major removal mechanisms from the atmosphere, but occult deposition by impaction of mist/fog appears to be important at higher altitudes.

The proportions deposited wet or dry vary with topography, climate and surface conditions; in many regions studied over time they are about equal.

Greater knowledge of the emission, transport and deposition of S and N pollutants in northern Europe and North America has led to the development of predictive models; as yet they have poor accuracy but can be used to develop further understanding.

Changes in the emissions of N and S over recent decades have been reflected in deposition chemistry at some sites, but there has been little overall change in the acidity of rain.

Acid rain in the strict sense must be distinguished from overall air pollution— the sources and targets may be different, calling for different control strategies.

3

The chemistry of deposition

3.1 THE NATURE OF RAIN

Although such a familiar aspect of our lives in temperate countries, rain is surprisingly difficult to define or characterize chemically. First, as explained in Chapter 2, for practical purposes *bulk* rainfall is collected in open dish-type collectors and comprises wet and some dry deposition. Secondly, sample collectors (although sometimes standardized) vary in geometry or height position, and in their siting—all factors leading to different efficiencies of collection, and to a different potential for contamination, e.g. from resuspended soil materials or bird droppings. Thus, the amount and quality of rain differ according to the precise siting of the sampler in relation to topography, altitude and wind exposure. It also varies in relation to antecedent conditions in the atmosphere which may not be known. Samples are rarely collected on the basis of each shower, but rather on some regular timed frequency, such as that accumulated during daily or longer intervals—a practical procedure but not consistent with the essentially stochastic nature of rain events. Evaporation from the accumulated sample, or occasional loss of some of the collected sample, may also occur during heavy rain spells. Finally, analysis is difficult due to the very dilute nature of the sample, especially with regard to acidity (often only a few μeq l^{-1}) and the instability of nitrate and ammonium ions, which are important constituents of the ion balance even at low concentration. Some methods of analysis commonly used are imprecise at these concentrations; even sulphate analyses in quite recent investigations lacked precision. Cooperative national and international programmes have tackled this problem with success, specifying both sampling and analytical procedures and carrying out 'blind' testing of samples at candidate laboratories. Individual investigations reported in the literature have thus to be accepted with appropriate reservations.

These problems are partially overcome if the objective of an observation is to compare rain chemistry at a single site over a period of time, following a consistent procedure. Unfortunately, rather few data sets are available to provide a valid comparison for differences in rain chemistry between regions or periods, especially where methods of both sampling and analysis have changed over time. Many older records were derived from infrequent sampling, where contamination was not controlled, or where subsequent analysis was delayed, as well as being imprecise by today's standards. Systematic sampling was seldom achieved prior to the mid-1950s, so long-term trends are not established. Thus historical or regional comparisons have to be accepted with caution—the claim that rain *has* become significantly more acid in the last 50 years, as discussed in Chapter 1, may not be soundly based although it is widely accepted.

Deposition is calculated on the basis of the measured coincident rainfall and its chemistry, usually of bulk deposition, as defined above, but will underestimate the total. In the United Kingdom, the standard rain sampler is not considered a precise gauge of rain quantity, and volumes are usually measured with a tipping bucket system, recording the time of each tip (0.2 mm) collected; this record is then converted to the total collected in unit time, e.g. over 24 hours. Unfortunately the distribution of the Meteorological Office solid state event recorders (SSER) in the UK may not correspond with the sites being investigated, but at some sites a good measurement of daily rainfall can be achieved. Given a reasonably precise measure of the rainfall quantity, **bulk deposition** to unit area (m^2) is then calculated from the product of the volume and ion concentrations.

The contributions of snow, mist and fog (**occult deposition**) are not adequately sampled by standard rain collectors. Some recent studies have addressed this problem, recording significant differences in volume and chemistry of rain at different altitudes and with different plant cover (Fowler *et al.*, 1989). Cloud deposition is estimated to account for 30% of total deposition of sulphur and nitrogen in areas of upland (>800m) Britain (Fowler *et al.*, 1991), adding about another 10% of sulphur and nitrogen to wet deposition over forest. In the northeastern USA, concentrations of sulphur and nitrogen species are 5 to 20 times larger in cloud water than in precipitation at high-altitude sites; droplet deposition, also higher at higher elevation, may be up to three times greater there, although these conditions apply to only a small fraction of the total site area (Irving, 1991).

Wet deposition in the form of snow presents another problem in its collection, since standard wet deposit gauges are inadequate. At some studied sites at altitude or at inland continental sites this may constitute a significant fraction of annual wet deposit (a third or more at Wood's Lake, Adirondacks: Johannes *et al.*, 1984), bringing significant acid episodes to stream water during melting (see below, and Chapter 6).

For practical reasons, budgets of catchment input are calculated on the basis of bulk rain samples, usually sited in adjacent open terrain. The errors in such calculations vary for a variety of reasons, as explained above. The flux of acid from dry deposition from air with a concentration of about 5 μg SO_2 m^{-2} (but typically less at clean rural sites) is estimated as equivalent to that in 1000 mm rain with pH 4.3–4.6. Away from point sources or conurbations the dry and wet components of deposition are thought to be about equal. Where there is heavy rainfall, as in the western seaboard of Scandinavia and Britain, wet deposition is likely to be greater, partly because of the frequency and quantity of rainfall, and because the atmosphere is so often 'scrubbed' of its atmospheric burden. It is in these high-rainfall areas that soil and surface water acidification is more prevalent, not so much because of high rain acidity (i.e. concentration) but rather because of the high level of acid *flux*. Effects are further enhanced by the rapid runoff and the progressively depleted soil base reserves. Rain reaching the soil is significantly altered by the pathway it follows through vegetation with its higher 'catch' of droplet materials. It is this **throughfall** material which is modified further by chemical changes within the soil and transferred via runoff to surface waters.

The current approach to emission control of sulphur and nitrogen emissions adopts the concept of **critical loads**, i.e. levels of deposition which will not cause chemical changes leading to long-term harmful effects on the most sensitive ecosystems (Nilsson and Grennfelt, 1988). This concept will be discussed in relation to vegetation, soils and surface waters in the following chapters. However, it is worth noting here that *total* deposition of sulphur and nitrogen is probably underestimated at most sites, possibly excluding urban sites (Irving, 1991).

3.2   RAIN CHEMISTRY

Since the composition of rain reflects atmospheric and climatic conditions (Fig. 3.1), it is not unexpected to find that it differs between sites, and from year to year, even from shower to shower. It also follows that rain sampled in polluted sites will differ from that at 'pristine' sites (Table 3.1). At polluted sites where acidity is often greatest, there are usually higher concentrations of sulphate and nitrate than at 'pristine' sites, although the relationship between regional sulphur emissions and rain sulphate is non-linear, and sulphur deposition is not dominated by nearby source regions, due to the variety of sources, some far afield, and the limitations imposed by oxidant availability (see Chapter 2).

Unneutralized sulphate has been regarded as the principal agent of rain acidity, but the nitrogen species (NO_x and NO_3^-), ammonium (NH_4^+), bicarbonate (HCO_3^-) and chloride (Cl^-) are also present. The latter two are almost always ignored, but the nitrogen species are certainly significant at pristine sites, and chloride may be high in industrial areas. The range of sulphate concentration in rain varies more than tenfold, from about 3–11 μeq SO_4^{2-} l^{-1} at pristine sites (San Carlos, Venezuela, and Amsterdam Island in the Indian Ocean: Galloway *et al.*, 1982), up to 160 μeq l^{-1} at Solling in western Germany (Ulrich *et al.*, 1979). Typical sulphate concentrations in urban and industrialized areas of Europe are above 50 μeq l^{-1} for sulphate, about 30 μeq l^{-1} for nitrate and about 15 μeq l^{-1} for chloride (all figures as non-sea-salt ions). Sulphate concentrations in rain over the UK (range 27–120 μeq l^{-1}) are highest over the Midlands and Yorkshire, nitrate (range 3–63 μeq l^{-1}) highest over eastern England, and chloride (range 47–515 μeq l^{-1}) highest over Humberside (Irwin *et al.*, 1990). The degree of matching acidity is governed by the amount of neutralizing material in rain: in the UK the highest acidity (pH <4.1) is centred over the Trent valley. In the eastern USA, rain is often more acidic than in Europe, possibly due to lower concentrations of ammonium there (Irving, 1991). The relationship of nitrogen emissions to concentrations of nitrogen species in rain is less well quantified, but both historical and spatial trends confirm a recent increase in nitrate in rain (Fig. 3.2),

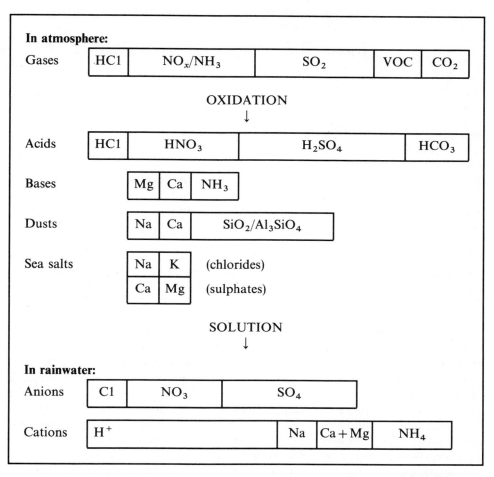

Fig. 3.1—Genesis of an acid rain water. Strong acids in rain are formed from acidic precursors in air reacting with bases (dusts and NH_3); the rain contains the dissolved anions and their associated strong acids (H^+) as well as ammonium (NH_4^+) and sea salt ions. The strong acids are formed by oxidation of atmospheric pollutants and react with ionized bases, dusts and sea salts. Acid rain has an excess of strong acids. (Modified from Stumm and Morgan, 1981.)

with higher values closer to sources. Ammonia (NH_3) and ammonium (NH_4^+) reduce rain acidity, but on deposition convert to nitrate (NO_3^-), and are taken up directly by foliar canopies, with an acidifying effect at the root zone (Nilsson *et al.*, 1982).

There is good historical information for the concentrations of nitrogen compounds in rain for over a century (Smith, 1872; Brimblecombe and Stedman, 1982). At Rothamsted (in southeast England) agricultural interest in the nitrogen nutrition of crops led to a time-series record from the 1850s through to the time of the First World War. The data can be reconstructed to show the seasonal and year-to-year

Table 3.1—Acidity and non-marine sulphate concentrations in bulk rain
at some remote and impacted sites

Location	pH mean/range	Sulphate (μeq l^{-1})	Comment
Amsterdam Is. (Indian Ocean)	4.92 4.0–5.5	11.5	Mid-ocean, remote
San Carlos (Venezuela)	4.69 4.0–6.7	3.0	Remote site
Solling Forest (Germany)	4.11	161	Impacted site
Woods Lake (Adirondacks)	4.2 3.4–5.8	57.4	Intermediate site, impacted
Hubbard Brook (NE USA)	4.14	60.3	Intermediate site, little impact
Tillingbourne (SE England)	4.15 3.4–6.1	80.5	Intermediate site, some impact
Loch Fleet (SW Scotland)	4.74 3.8–6.0	38.7	Fairly remote, impacted

changes in rain constituents, including nitrate and ammonia, for comparison with recent measurements across Europe (Fig. 3.3; Skeffington and Wilson, 1988). A clear seasonal variation can be seen in both historical and modern ammonia concentrations, with summer maxima reflecting both soil and litter production as well as livestock and industrial emissions. Nitrate, in contrast, showed little seasonal variation in the past, but now has spring and summer maxima and a doubling of total deposited since about 1900 (Brimblecombe and Stedman, 1982), although rainfall quality is not significantly different between the two periods (Fowler and Brimblecombe, 1988). Similar increases are seen from US records since 1890.

A further significant influence on rain chemistry in maritime areas comes from the contribution of sea salts entrained in the atmosphere. Much of the sulphate and other ions washed out of the atmosphere by rain in the UK is derived from sea salts; this was recognized by Eville Gorham in the mid-1950s, since rain samples he collected in the English Lake District reflected season and the predominance of rain from the southwest, characterized by Na:Mg:Cl ratios similar to those of sea water (Gorham, 1955). This marine component is generally assumed to be neutral and not affecting its acidity or acidifying potential (Brydges and Summers, 1990). However, sea salts contribute to the ionic strength of the rain and reduce its acidity since they are somewhat alkaline (pH 8.25). They also have significant influence on the release of acidity from acidified soils (see Chapter 5). Another point is that the *total* input of sulphate deposited to soils adds to the soil sulphur reserve, where its retention is considered to be an index of 'stored acidity', reducing the expected benefit of reduced sulphur emissions from industry (see Chapters 5 and 11).

Snow, originating from cloud, has initially a chemistry similar to rain. However,

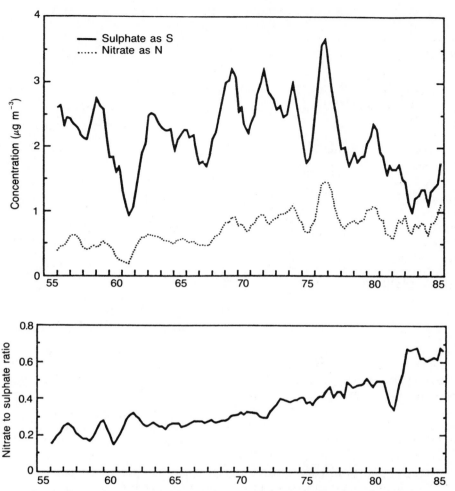

Fig. 3.2—Trends in sulphate and nitrate deposition since 1978 at a site in central England
(Barrett *et al.*, 1987).

during deposition it may scavenge significant amounts of dusts, enhancing levels of
trace metals. There is also the likelihood that fallen snow takes up dry deposited
material, especially if coincident with pollution episodes. At melting, snow solutes
become fractionated so that the initial melt water may be much more concentrated
than rain, or bulk snow. Further, if the underlying ground is frozen, there is little or
no opportunity for neutralization from soil contact (Johannessen and Henriksen,
1978). These processes lead to highly acid, sometimes toxic, episodes in snow-fed
streams (see Chapter 6).

A final atmospheric constituent, dust, is also often neglected; it is significantly
related to climatic extremes (drought and dust storms, common in the 1950s in the

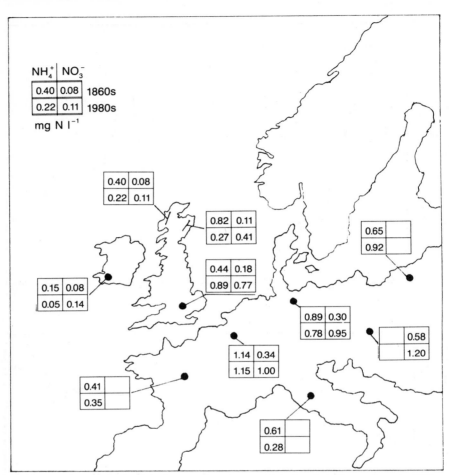

Fig. 3.3—Changes in nitrate and ammonium concentrations in rain in Europe between 1880 and 1980 (Skeffington and Wilson, 1988).

USA) and also to poorly controlled dust emissions from industry in earlier years (Hansen and Hidy, 1981). It is certain that the chemical nature of rain is determined by its *total* ionic composition, not by a selection of ions. Many records are inevitably flawed by lack of measurement of many significant components, and good records for past and present levels of dusts, calcium and magnesium carbonates, as well as sulphur and nitrogen species, are needed for convincing evidence of a worsening trend in rain acidity (see Chapter 1).

So, notwithstanding variations in rain composition which result from short- or long-term changes in climate, rain varies according to location and distance from sources of acid-generating pollutants. Major ions of interest are, of course, hydrogen ions (H^+), ammonium (NH_4^+), sulphate (SO_4^{2-}) and nitrate (NO_3^-), but in coastal regions sea salt ions may dominate: sodium (Na^+), chloride (Cl^-) and also magnesium

Table 3.2—Rain composition at some impacted sites in Europe and North America and at some remote sites

Site	$[SO_4^{2-}]$	$[NO_3^-]$	$[NH_4^+]$	$[HCO_3^-]$	$[H^+]$
Amsterdam Island	29.2	1.6	2.4	2.1	8.3
Amsterdam Island*	4.8	1.6	2.4	2.1	8.3
Katharine, Australia	3.9	4.0	2.9	7.3	18.3
Lijiang, China	5.3	1.6	4.7	3.4	9.7
Gardsjon, Sweden	35.0	34.0	50.0		50.0
Birkenes, Norway	46.0	30.0	33.0		42.0
Loch Fleet, UK	48.9	17.8	26.2		22.2
Woods Lake, USA	57.4	33.1	19.4		59.5

Concentrations are $\mu eq\, l^{-1}$; *for non-marine fraction.
Source material: Birkenes, Henriksen *et al.*, 1988; Gardsjon, Grennfelt *et al.*, 1988; Fleet, Howells and Dalziel, 1991; Amsterdam Island, Katharine and Lijiang, Hicks and Artz, 1992.

(Mg^{2+}). In rural and desert areas, dusts may be significant, for example near cement or lime works, or where agricultural ploughing and burning are practised. Some contrasting examples of rain composition at different locations are given in Table 3.2.

A notable difference is seen in the characteristics of rain in Europe and the USA. While rain in the two regions is equally acidic, the total ionic strength of European rain is two to three times greater than in that falling in the USA, attributable to a higher content of ammonia, calcium and magnesium in European rain (Verry and Nodop, 1992). Western coastal areas of Europe receive a fourfold larger input of sea salt ions in rain than do areas inland.

3.3 CHEMISTRY OF FOG, MIST AND SNOW

The different nature of rain collected in open samplers at low altitude and of that collected at higher altitude has been introduced earlier (see Chapter 1). Further recent studies on this 'occult deposition' have provided more evidence that cloud, mist and fog water have significantly higher concentrations of major ions (Crane and Cocks, 1987; Fowler *et al.*, 1991). These observations are consistent with samples of cloud water collected in the atmosphere. This has led to suggestions that the deposition of gases and the impaction of water droplets from low clouds could be a significant and direct pathway from the atmosphere to terrestrial targets, particularly at high altitudes.

Studies in the UK at Great Dun Fell (847 m), in western Germany and in the USA at high-altitude sites (> 1400 m) have demonstrated the importance of this pathway, albeit limited in areal terms. At such sites in Britain (Fowler *et al.*, 1991; Irving, 1991; Schmitt, 1988), in Germany and the northeastern USA, low cloud is present during significant periods (30–40% of the growing season in the USA). The ratios of major ion concentrations in cloud to those in rain often show a threefold difference. Cloud water samples analysed for several sites are illustrated in Table 3.3 and may be compared with those shown in Table 3.2. Collection of deposition at the canopy level confirms that it is higher, although possibly less important to the target vegetation

Table 3.3—Concentrations of ions in cloud water at some upland sites in the UK, the USA and Germany

Site	Altitude (m a.s.l.)	$[H^+]$	$[SO_4^{2-}]$	$[NO_3^-]$	$[NH_4^+]$
Smoky Mts, USA	1800	250	350	80	140
Great Dun Fell, UK	840	151	221	140	250
Kleiner Feldberg, Germany	800	199	255	164	210

at these sites, than cloud water composition since direct gas exchanges (especially highly reactive NO_x, HNO_3 and HCl) will also occur. The high rates of ammonium deposition (20–60 kg N ha^{-1}) (Fowler *et al.*, 1991) must be a significant nutrient source in upland areas, and co-deposition of ammonia and sulphur dioxide within canopies also enhances sulphur deposition. It is claimed that an enhanced 'deposition velocity' of atmospheric constituents adds about 20% to the bulk deposition at altitudes > 400 m (Dollard *et al.*, 1983; Barrett *et al.*, 1987), but it is difficult to confirm this by field measurements, or to judge its importance in relation to catchment responses. The enhanced sulphate deposition is probably only a few percent for upland catchments, but could be more significant for small, high-altitude, forested sites.

It should be noted, notwithstanding the lack of inclusion of occult deposit, that reasonable input/output budgets for catchments can be calculated using bulk deposition, estimated dry deposition, evapotranspiration and some soil retention (e.g. for Gardsjon in southwest Sweden: Hultberg, 1985; Hultberg and Grennfelt, 1992), suggesting that any additional input from occult deposition is not very significant.

The contribution of snow to runoff chemistry is even more difficult to quantify. Clearly, snow composition reflects antecedent atmospheric conditions and concentrations, as well as subsequent pollution events that may have contributed to additional deposit, e.g. of 'soots' (Tranter *et al.*, 1988) following a combustion incident upwind. Further, the fractionation of snowmelt water during successive melt and thaw cycles delivers a variable load of acidity or toxic agents to runoff. Laboratory and field studies of this process in Norway suggested that about 30% of melt could contribute 50–80% of the snowpack pollutants (Johannessen and Henriksen, 1978). Samples of snow are seldom reported with coincident rain composition; some indication of their relative acid concentrations may be gathered from the data in Table 3.4.

3.4   CHEMISTRY OF THROUGHFALL

Both dry deposition and occult deposition are intercepted by canopies which generally enhance solute concentrations in the throughfall. Data for throughfall below tree canopies, often for isolated trees or those at the margins of stands, demonstrate that such scavenging of the atmosphere occurs (Grennfelt and Hultberg, 1986). Some examples are shown in Table 3.5, compared with bulk rainfall; some measurements

Table 3.4—Composition of rain and snow at some monitored sites in Europe and North America

Sample	pH	$[SO_4^{2-}]$	$[NO_3^-]$	$[NH_4^+]$	$[Cl^-]$	$[Mg^{2+} + Ca^{2+}]$
UK, rain:						
(Average)	5.35	68	29	30	25	19
Cumbria	4.5	60	22	28	124	51
E. Scotland	4.7	30	11	6	66	23
UK, snow:						
Gt Dun Fell	3.8	32–73	14–26		140–210	
E. Scotland	4.7	53	24			
Norway, Birkenes:						
Rain	4.2	46	30	33	52	17
Snow	4.2	180	40			
Sweden, Gardsjon:						
Rain	4.3	35	34	50	65	10
Ontario, Canada:						
Rain	4.2	48	34	19	4	10
Snow	5.0	110	15			
Nova Scotia, Canada:						
Rain	4.4	31	15	7	33	11
Snow	3.8	70				
NE America:						
Rain	4.1	60			14	11
Snow	4.4	100	50			

Ion concentrations are μeq l^{-1}.
Data from Brydges and Summers, 1989; Henriksen *et al.*, 1988; Grennfelt *et al.*, 1985; Irwin *et al.*, 1990; Ogden, 1982; Tranter *et al.*, 1988.

and estimates relate to forest stands (Novo *et al.*, 1992; Miller, 1984; Grennfelt *et al.*, 1985).

Throughfall samples have typically higher sulphate and nitrate concentrations as well as enhanced potassium and sea salts; they are usually higher in acidity, although this varies significantly between tree species (Nisbet and Nisbet, 1992; Draaijers *et al.*, 1992). These effects appear less important on a catchment scale, perhaps because the contribution of 'free fall' rain is predominant.

It is evident from these throughfall samples that rain collected in standard collectors in the open differs significantly in quality and often in quantity from that collected below trees (Miller, 1984; Bredemeier, 1988). This arises from a variety of mechanisms (Fig. 3.4). First, dry deposited material accumulates between rain events and accumulates on leaf surfaces; it is 'wetted up' by rain or fog, increasing the concentration of not only ionized soluble materials but also others which are less

Table 3.5—Comparison of fluxes from bulk deposition and throughfall
at some European sites

Site and conditions	Tree and age (yr)	Bulk deposition			Throughfall		
		H^+	SO_4^{2-}	Ca^{2+}	H^+	SO_4^{2-}	Ca^{2+}
Luneberg,	Oak, 105	0.8	23.4	10	1.3	50.3	24.1
500 m a.s.l.,	Pine, 100	0.8	23.4	10	3.1	85.2	32.5
800 mm rain, pH 4.26							
Solling,	Beech, 138	0.4	16.9	5	0.5	28.6	12.8
500 m a.s.l.,	Spruce, 103	0.4	16.9	5	1.0	36.0	17.4
1030 mm rain, pH 4.11							
Harz,	Spruce, 40	0.3	22.4	6	0.4	42.7	19.8
700 m a.s.l.,							
1270 mm rain, pH 4.17							
Mottarone,	Spruce, 50	0.8	24.8	10	1.2	40.3	26.8
900 m a.s.l.,							
1896 mm rain, pH 4.36							
Loch Fleet,	Spruce, 12	0.6	18.2	5	1.1	40.6	17.4
340 m a.s.l.,	Lodgepole	0.6	18.2	5	0.9	26.1	10.9
2200 mm rain, pH 4.75							

Values for ion deposition are for annual flux in kg ha^{-1}.
Data from Bredemeier, 1988; Nisbet and Nisbet, 1992; Novo et al., 1992.

easily washed out of the atmosphere by rain, including impacted aerosols and dusts.
It also mobilizes materials scavenged by canopies from mist and fog in the absence
of perceptible rain. Some atmospheric and rain constituents, like ammonia and
nitrate, are also directly absorbed by leaf surfaces (foliar feeding), particularly where
the soils are deficient in nutrients (Nisbet and Nisbet, 1992). In contrast, some
materials are leached from leaf tissues to be washed off by falling rain, evidenced by
a higher concentration of potassium. In addition, some acidity may be neutralized,
in situ, by cation exchange at the leaf surface. A major change occurs by the
mechanism of evapotranspiration, which will concentrate all constituents by loss of
water from the leaf surfaces. Evapotranspiration varies with leaf physiology and plant
structure—tree canopies with their complex geometry lose much more water than
pasture or moorland (about 30% compared with 20%: Newson, 1975; Hall, 1987); it
is also affected by climatic and diurnal conditions, turbulence at the forest edge and
increases with high winds in dry air during the growing season. A greater volume of
throughfall is found below Douglas fir, less below Scots pine, and less again below
oak (Draaijers et al., 1992); throughfall deposition was found to be strongly related
to the crown volume.

These differences in chemical composition are illustrated in Fig. 3.5 (Nisbet and
Nisbet, 1992). The greater volume and concentrations in forest-intercepted rain
are reduced somewhat for some ions and acidity in transfer to runoff, due to
evapotranspiration and exchange processes. At Loch Fleet (with magnesium-deficient
soils), the input of sea salt ions is high, and sodium and chloride exceed sulphate in

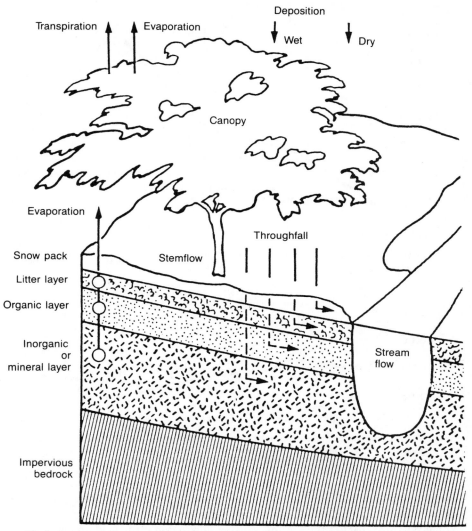

Fig. 3.4—Interactions of rain and dry deposit within a forest canopy—dry deposited materials accumulate within the canopy, to be washed off when rain falls. Some substances are taken up within the canopy (e.g. NH_4^+), while others are leached (e.g. K^+).

both concentration and flux. Nitrogen and magnesium are taken up in the canopy, and potassium leached. It is notable that large gains in potassium in the throughfall during summer and autumn (possibly by exchange with H^+ in the canopy) did not lead to a reduction in the H^+ flux, thus failing to neutralize some of the incoming acidity. This has been interpreted (Nisbet and Nisbet, 1992) as evidence for the transfer of neutral K_2SO_4 via canopy leaching, while base cation exchange was small. In other areas with richer soil minerals these relationships are likely to be different.

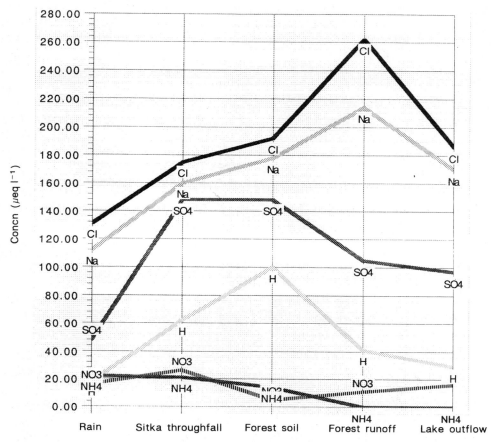

Fig. 3.5—Comparisons of rain, throughfall, soil percolate, runoff and lake water concentrations, and of their fluxes at a forest (spruce) site at Loch Fleet, southwest Scotland (data from Nisbet and Nisbet, 1992).

A problem also lies in distinguishing these various processes quantitatively since they cannot be measured independently of one another.

Some rain also reaches the ground via **stemflow**, again more acid and concentrated than incident rain. The limited data available allow little generalization, especially on the volumes generated in different conditions. It is clear that volumes are highly variable, not only between tree height and size, but also with rain intensity (Miller, 1984); at Loch Fleet in Scotland stemflow was 14–17% of bulk precipitation input below spruce, but twice this below lodgepole pine (Nisbet and Nisbet, 1992). Even though the volume is so much smaller than that of throughfall, it may contribute up to as much as half the influx of acidity reaching the soil below a forest. It has been suggested that this may be the cause of the typical podzolization of soils close to trunks of trees, although the activity of the root biomass may also play a part in extracting soil cations (Skeffington, 1983).

At a number of well-studied sites in Germany, the mean pH of throughfall (3.4 to 4.7) was typically lower than that of contemporary bulk precipitation (pH 4.1 to 4.6) (Bredemeier, 1988); a similar pH reduction was seen in a spruce forest in the Prealps (Novo et al., 1992). At some other sites in northeast Scotland, pH in throughfall was higher than in bulk rain (Miller, 1984). Such differences reflect canopy characteristics and leaf activity (Draaijers et al., 1992), as well as variations in the dry deposit contribution at different sites. Indeed, the use of throughfall rather than bulk deposition is arguably a better measure of total deposition (Hultberg and Grennfelt, 1992).

The quantitative evaluation of deposition below trees is difficult without more information from a variety of sites. More studies at metered catchments are needed so as to estimate the relative proportions of throughfall and stemflow reaching the ground, as demonstrated at the afforested Gardsjon (Sweden) catchments, where 67% of deposition comes as throughfall, while mist and fog interception adds a further 13%, only the remaining 20% coming via freefalling rain (Grennfelt et al., 1985; Hultberg and Grennfelt, 1992). The transfer of throughfall, stemflow and freefall rain is subsequently modified by its pathways, interactions and rates of transfer through the litter and soils of the catchment (see Chapter 7).

The mineral and nutrient status of the trees is a significant factor influencing the composition of throughfall. There is good evidence that throughfall below N-deficient trees is much more acid than the incident bulk rain, and that application of a suitable fertilizer reduces the acidity of throughfall and is accompanied by uptake of ammonium by the canopy (Miller, 1984; Nilsson et al., 1982; Nisbet and Nisbet, 1992). Cation exchange at the leaf surface can exchange H^+ for calcium or magnesium, so that Ca^{2+} and Mg^{2+} are increased; potassium and manganese are also leached (Nisbet and Nisbet, 1992; Fritsche, 1992). Recent laboratory findings (Fowler, 1992) indicate that SO_2 deposition to wet canopies is promoted by the presence of NH_3, since this neutralizes the acidity generated by the solution and oxidation of SO_2. Foliar leaching of sulphate was estimated by Parker (1983) to be as much as half the sulphate in throughfall. Richter et al. (1983) concluded, however, that a large part of the excess sulphate in throughfall (about 50% from annual fluxes) was derived from atmosphere, but tracer studies have now demonstrated that some foliar leaching does occur (Cape et al., 1992), providing a route by which soil sulphur is cycled through the canopy. In a four-month study in a Scots pine forest, about 3% of net throughfall could be attributed to this process, assuming that the tracer S is equilibrated readily in the soil.

Thus, throughfall composition is a consequence of several interactive processes occurring at the canopy level. Interception of atmospheric constituents via dry, wet, or occult processes increases the content of many solutes, while leaf leaching contributes additional materials. Intermittent rain showers, or mist and fog conditions provide for the dissolution or entrainment of dry or occult deposited material between rain events. In addition the enhanced evaporation from forest canopies (about 10% greater than from moorland or grassland in UK conditions: Hall, 1987) concentrates the solutes in throughfall. While these phenomena are mostly demonstrated by studies below forest canopies, the same processes are found for other vegetation, such as shrubby heather moorland, although to a lesser extent (Nisbet and Nisbet, 1992).

3.5   SEASONAL AND SHORT-TERM EVENTS

The operation of several regional and national networks of rain sampling has provided information about the short-term and seasonal variations of rain quality. Some of these variations relate to the interaction of climate variables with site topography, and their interpretation is important in deducing longer-term trends that might be attributed to changes in acidifying sources.

Climatic conditions often result in large rainfalls, sometimes preceded by extended dry periods. In these conditions, the atmospheric burden of acid-generating pollutants may have been accumulated so that initial rain-out scavenges high atmospheric concentrations of oxidized materials. The antecedent dry period may also have reduced soil permeability, so that any rainfall may scarcely penetrate the soil, with superficial and rapid runoff to drainage channels. This allows little interaction between the incident rain and soils, with immediate runoff composition close to that of incident rain. Such events may be associated with very low pH levels in rain, possibly corrected with time as soil conditions improve, allowing neutralization by base mobilization from soils, or by mixing with baseflow water which has been in contact with basic soil and mineral material. Seasonal conditions also affect the episodic nature of hydrological response to rain events, with dry conditions occurring in summer, and also in winter when the catchment is frozen or when accumulated snowfall begins to melt. Thus the response of upland streams to rain events is critically dependent on rain duration and intensity, soil moisture and infiltration capacity, and on hydraulic conductivity and thickness which determines the pathway of drainage flows through to streams.

In many places the amount of acidity deposited in a year comes in relatively few days. In the United Kingdom, for example, at one southwestern site 30% of total acidity was deposited in 2.7% of wet days (5 of a total 185 days) (Barrett et al., 1983); sulphate and nitrate deposition showed the same characteristics. At principal UK national rain monitoring sites, the days of rain accounting for 30% of deposition range for acidity from 2.3 to 5.9, for sulphate from 2.8 to 8.1, and for nitrate from 3.1 to 6.3 (Irwin et al., 1990). Rainfall days at Loch Dee in southwest Scotland are more frequent, and events of low pH and high sea salts are also frequent (15 and 20 days respectively in 1990) (Lees and Farley, 1993). During one such event rain acidity and conductivity showed variations in the range 5–50 μeq l^{-1} and 5–160 μs cm^{-1}; this was associated with passage of a frontal system, bringing high-intensity precipitation over a period of 9 hours. Later analysis showed high sodium (422 μeq l^{-1}) and chloride (471 μeq l^{-1}) concentrations, accounting for the high conductivity (Langan, 1987, 1989). The stream flows following the heavy rain showed an almost instant response, with a transient increase in acidity and conductivity (Fig. 3.6). Similar natural events with high levels of sea salts are also reported for Norway with two to three occasions a year with chloride > 150 μeq l^{-1} (Wright et al., 1988), and extreme acid events causing fish kills (see Chapter 9). This response was also demonstrated by a field sea salt application; chloride concentrations increased from 42 to 302 μeq l^{-1}. The consequences for stream waters are important, with sea salt events being associated with release of acidity and aluminium from soils (see Chapter 5).

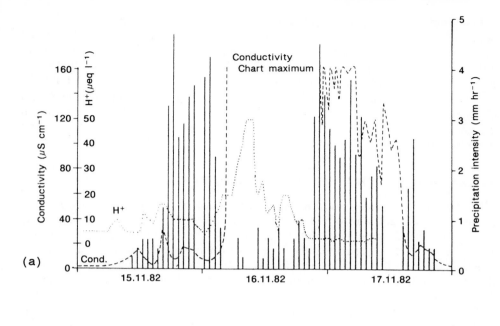

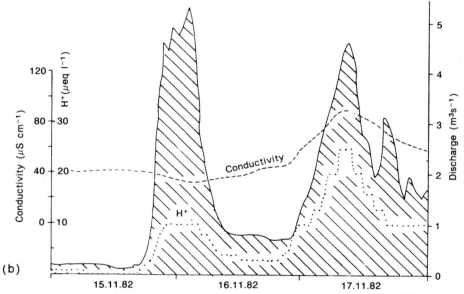

Fig. 3.6—Short-term events in (a) rain and (b) stream at Loch Dee, southwest Scotland (after Langan, 1987, 1989); hourly values during two episodes in 1982. The increased conductivity values in rain were associated with sea salt sodium.

On a somewhat longer timescale, a recent increase in rain pH and a fall in sulphate concentration since 1982 observed at sites in western Germany has been attributed to three mild winters when there was significantly less effect of an easterly air flow, although there was also a concurrent fall in the emissions of sulphur (Fricke and Beilke, 1992). Trends in atmospheric circulation over Europe are known to influence wet deposition patterns over timescales of years to decades (Davies et al., 1992), with changes in zonal pressure gradients explaining up to 60% and 46% of the annual variance in nitrogen and sulphur components respectively in rain in the United Kingdom. For individual months, the proportion can be $>90\%$.

3.6   REGIONAL AND TEMPORAL TRENDS IN RAIN AND DEPOSITION

Since man-made emissions are a significant source of constituents of rain in the northern hemisphere, it is to be expected that changes in rain quality will be broadly parallel to the development and operation of industry in developed countries. However, such trends are quite difficult to establish for specific sites or areas, possibly due to inadequate historical records for rain; they are even uncertain for the past few decades in spite of improvements in sampling and analytical techniques (Galloway and Rodhe, 1991).

Analyses of rain in Britain were started as long ago as the 1820s, when the observations of the chemist John Dalton led to the conclusion that the trajectory of air masses strongly influenced their composition (Fowler and Brimblecombe, 1988); thus rain from the southwest was marine in character, while that originating from the industrial north was high in sulphur. In the industrial city of Manchester, Smith (1852, 1872) was able to demonstrate that concentrations of sulphur dioxide in the air were very high (1800 μg m^{-3}) and that rain acidity in urban areas was an order of magnitude higher than in rural locations and was associated with sulphate. At about the same time, the possibility that atmospheric nitrogen could fertilize agricultural production led to a long-term programme of sampling and analysis at Rothamsted (southern England) that was to continue for nearly a century. In 1897, emissions of soot in Leeds were estimated at 20 tons a day, of which half was deposited in an area of 4 square miles; a later account of sampling at 10 sites in the city in 1909 found nearly 2000 lb acre^{-1} (0.225 kg m^{-2}) of impurities at the centre (Cohen and Rushton, 1909). Although most of this particulate material deposited near to its sources, some undoubtedly travelled further afield, leading to observations such as those of Brogger in 1881 that 'dirty snowfall' in Norway could be attributed to industrial sources in Britain.

Early in the present century fish kills were observed in some Norwegian rivers and mountain lakes and current developments in chemical techniques were able to link this with pH values in surface waters of <5.0 (Strom, 1925; Dahl, 1927), although the cause was only retrospectively attributed to episodes of pollution originating from Britain (Dannevig, 1959; Oden, 1967).

At this time, Gorham (1958) also reaffirmed that rain quality reflected the proximity and upwind source of acidic gaseous emissions, or alternatively in the UK of maritime influences. Swedish scientists also rekindled interest in the potential of rain for crop nutrition, as well as its potential for adverse effects due to acidity. This led to the establishment of the European Air Chemistry Network (EACN) in 1947. Initially established in Sweden, its scope was later widened to include stations in 11 European countries, and interest grew in constituents other than sulphur. The total of collaborating stations was at a maximum (about 120) in 1959; major findings are reported by Granat (1978). In spite of early observations of a large increase in rain acidity during the 1960s, over the longer term to 1975 most stations showed a trend in increasing sulphate concentrations but also a large year-to-year fluctuation in both sulphate and rain quantity, reducing the changes to statistical non-significance (Kallend *et al.*, 1983).

Interpretation of these records has varied, according to the selection of the period of observation and the sites selected, sometimes perhaps to justify such a trend. For the period from 1955 to 1975, Oden (1976) and Likens *et al.* (1979) used data from Norwegian and Swedish sites to show a four- to ten-fold increase in rain acidity; Soderland (1977) found a greater than twofold increase in nitrate deposition. Re-evaluation of these data by Kallend *et al.* (1983) showed that for the 120 stations with data for this period, a positive and statistically significant trend in concentrations was evident at only 29 sites for acidity, at 23 for sulphate, and at 55 for nitrate. Five stations showed a decrease in acidity and only at 10 stations was there a concurrent trend in both sulphate and acidity, while at 18 there was a parallel trend in nitrate and acidity. A change in ammonium concentration and deposition was even less clear, but analyses are flawed by the possibility of bacterial contamination of samples awaiting analysis. A relationship between trends in sulphur emissions and deposition is thus not convincing—though sulphur emissions increased, peaking in 1975 in Europe, deposition over the same period decreased. Differences between sampling sites suggest that local conditions of both emission and deposition mask any consistent and overall trend.

A more recent attempt to clarify the regional and temporal pattern of rainfall over Europe was sponsored by the OECD (Organization for Economic Cooperation and Development); this began in 1973 and involved 73 stations. The deposition of 'excess' (non-marine) sulphate estimated from the OECD stations confirmed the findings of the earlier EACN work, but better experimental design, station siting and analysis methods resulted in less year-to-year variance. This programme grew into the current cooperative UNECE (United Nations Economic Commission for Europe) programme for the monitoring and assessment of long-range transmission (European Monitoring and Evaluation Programme, EMEP), which has, since 1977, taken over the role of measuring and modelling emissions and deposition of acid pollutants.

Granat and others showed that from 1955 to 1975, an increasing area of Europe received more than background levels of sulphur deposition; a similar extension was found in the northeastern USA (Atshuller and Linthurst, 1984). However, since that time there has been a contracting deposition field in Europe; in only a small part of the monitored area, sulphate deposition increased in 1965–1975 when sulphur emissions increased by 35% (Goldsmith *et al.*, 1984), while elsewhere it remained constant or declined. A recent analysis of trends in Sweden (Sanden *et al.*, 1987) shows that from 1969 to 1982 when sulphur emissions were reduced twofold, at only

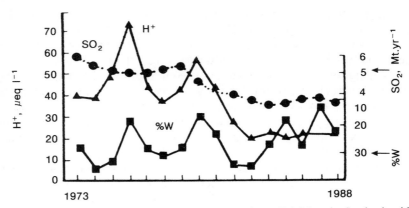

Fig. 3.7—Comparison of annual mean acidity of rain at Eskdalemuir, Scotland, with national SO_2 emissions and the frequency of westerly rainfall over a 15-year period (after Irwin *et al.*, 1990).

three of 15 data sets could a change in acidity be statistically verified. This can be reasonably attributed to long-term variations which differ between areas. Similar modelled isopleth maps of deposition have been published for the USA (Likens *et al.*, 1979; Alsthuller and Linthurst, 1984) but limited data and timespan of recording make it still difficult to assess deposition response to changes in regional emissions (Irving, 1991).

In the United Kingdom, a 'primary' network of nine 'wet-only' monitoring sites has been established with samples analysed on a daily basis; a further 59 'secondary' network sites provide data from bulk samples on a weekly or fortnightly frequency (Irwin *et al.*, 1990). Annual mean concentrations at one remote site (Eskdalemuir) of acidity and westerly rain over the years 1973–88 are shown in Fig. 3.7. Considerable variation is found between sites—for some Scottish sites, there is some correlation between emission changes and decreased acidity and sulphate over the period 1980–85 (Harriman and Wells, 1987), but little correspondence between the two. Statistical analysis using autocorrelation, linear regression or Spearman rank correlation techniques found significance in the reported changes of sulphate only at Eskdalemuir, although all northern sites together showed a decrease in acidity ($p < 0.0003$) (Irwin *et al.*, 1990). The same trend is not evident at more southerly sites, closer to major emission sources.

It has to be concluded that the variability between sites and years is so large, and the duration of recording so short, that significant temporal trends do not match the 40% reduction in sulphur emissions achieved up to 1988. The results are consistent with the view that only part of deposition at sites distant from sources is attributable to national emission (Barrett *et al.*, 1987). A major contributing factor is the variability of weather—application of the EMEP model demonstrated that the observed fall in deposition could be attributed to weather alone (but was not statistically significant). There is perhaps an explanation in the finding of a greater frequency of westerly rainfall at Eskdalemuir in the last decade (Irwin *et al.*, 1990) (Fig. 3.7).

In North America, until recently, rain has not been monitored as intensively as in Europe. Junge and Werby (1958) analysed monthly samples from 17 stations in the eastern USA during 1955 and 1956. Their measurements did not include pH but this was subsequently estimated from ion balance by Cogbill and Likens (1974); pH ranged from 4.45 in Vermont to 6.0 in eastern Carolina. A follow-up survey was undertaken between 1965 and 1966, 10 of the stations being at Junge and Werby's sites. A similar range of pH was reported, from 4.2 in New Hampshire to 6.04 in northwest Carolina. A further survey was conducted 10 years later, at 16 stations of which eight were common with the 1962 study but only two with the original (1955) stations: one of these showed an increase in pH, the other a decrease! The current US programme, the NADP (National Atmospheric Deposition Program), is both more extensive and more intensive, with a weekly sampling schedule at 50 sites and good analytical control. This was incorporated into the US Environmental Protection Agency (EPA) NAPAP (National Acid Precipitation Assessment Program) which included 150 stations. Overall, the annual mean acidity of rain ranges from pH 4.1 in the northeastern USA to pH 4.5 in Florida, and to pH 5.1–5.6 in western states. Data for both the USA and Canada over the period 1979–87 showed a decrease in both sulphate concentrations (20%) and deposition (27%), but no significant trend in acidity. The areas of highest rainfall have, in general, the highest levels of sulphate and nitrate deposition, similar to the pattern in the UK.

Another data set comes from the Global Trends network (GTN)—locations selected on the basis of their remoteness from possible pollutant sources. Rain chemistry at these sites over the operating period suggests that 'pristine' or background pH in rain is about 5.0 (see Fig. 1.3) with excess sulphate concentrations 4–9 μeq l^{-1}; it is little changed. Some sites with pH $\geqslant$ 5.8 are influenced by dusts, wild fires and volcanic activity (Irving, 1991).

3.7   THE CONCEPT OF CRITICAL LOADS

In searching for a scientific strategy for control of acid-generating emissions, the concept of **critical load** (Nilsson, 1986; Nilsson and Grennfelt, 1988) has been accepted by UNECE. Definitions have varied according to the target response relevant in terrestrial and aquatic ecosystems, but incorporate the need to limit emissions to the 'highest load that will not cause chemical changes leading to long-term harmful effects on the most sensitive ecological systems' (Nilsson, 1986). The term 'load' refers to the quantity of a pollutant deposited, thus it is the product of concentration and total deposition from atmosphere to the ground; it will be expressed as a quantity per unit area. The critical deposition load can now be estimated more completely, since occult deposition pathways are now better documented (e.g. Crossley et al., 1992); however, measurement or estimation of total deposition still poses a problem. It is also true that while sulphur and some nitrogen loads are reasonably well quantified, many acidifying or neutralizing components are not so well reported. Some of the gaps are covered by modelling exercises relating emissions to deposition (Derwent et al., 1989), although these are not always validated by subsequent measurements.

The basic premise is that the critical load is based on a dose–response relationship, widely used in setting pollution criteria for a variety of contaminants. As such this is preferable to arbitrary concentration levels, specific air quality standards (e.g. US Clean Air Act, 1990) or the '30% strategy' (of 1980 sulphur emissions or transboundary fluxes) adopted by the UN in 1985 (UN, 1985), since there is no assurance that such strategies are sufficient or, for that matter, necessary. Indeed, a sceptical observer of the new US Clean Air provisions argues that they are designed to fail and are environmentally inefficient (Blake, 1991).

Difficulty is also experienced in applying the concept to atmospheric 'levels' (i.e. concentrations) of acid pollutants in the atmosphere which may lead to direct effects on targets such as vegetation or the terrestrial ecosystem. As with other pollution control strategies, the damaging concentrations have to be defined in terms of time, e.g. over 24 hours, or shorter or longer periods, as well as in relation to acute (short-term) or chronic (long-term) effects (Bull, 1991). These and other problems of defining target responses will be covered for vegetation, soils and surface waters in the following chapters.

3.8   UNRESOLVED ISSUES

After more than a century of rain sampling and measurement, and the intensive activity on both national and regional scales over the last two decades, it seems that many unresolved issues remain; recent investigations even seem to have increased these.

> What is the *true* level of total deposition at any site, and what is the best way to measure it?
>
> What are the quantitative relationships between emissions (of sulphur and nitrogen) and deposition?
>
> How will sulphate and nitrate levels in rain at specific sites respond to changes in emissions?
>
> Will oxidant limitation influence deposition if emissions are increased?
>
> Will increasing NO_x and VOC emissions counter any reduction in sulphur emissions?
>
> What are the local, regional or global patterns of deposition; what are the factors that explain differences between sites?
>
> In specifying air quality criteria or critical loads, what level of deposition will be specified for sensitive or other targets?
>
> To what extent should episodic events or the contribution of occult deposition be included?
>
> To what extent are geographical or temporal trends determined by emissions, climate variations, or by other factors?

3.9   SUMMARY AND CONCLUSIONS

Work on the chemistry and behaviour of deposition leads us to conclude

Sulphate is the major component of acidity in rain on a mass basis, but nitrate is increasing, and if ammonia nitrogen is included, their acidifying potential is similar to sulphate on an equivalent basis.

Rain composition varies widely between sites, reflecting climatic conditions, site characteristics and man-made source components.

Mist and fog are generally much more acidic than rain; at high-altitude sites this may enhance the acid load, although on an areal basis this effect may be small.

Throughfall and stemflow below tree canopies differ in composition from bulk rain; they vary according to species, the nutrient status and the complexity of the canopy. It is difficult to estimate the true atmospheric flux of acidity from throughfall.

Long-running records of rain chemistry are not often consistent with changes in sulphur and nitrogen emissions, although the match is better for nitrogen than for sulphur.

The confounding effects of climate, non-linear relationships, unknown background, and differences between sites and land use make it difficult to predict rain response to lower emissions.

The difficulties of maintaining a monitoring programme of sufficient geographical and temporal scale may also be one reason why expected trends are not clearly evident.

4

Plants and trees: acid gases and acid rain

4.1 INTRODUCTION

The effects of atmosphere on crops and on plants of conservation interest have been recognized at least since the middle of the seventeenth century. Research on the nature and extent of these effects, on their specific causes, and recently on plausible mechanisms or processes, has been, and still is, substantial. Notwithstanding this past and current effort, major gaps still exist in understanding and quantifying effects, especially where mixtures of pollutants or where other stresses are present, and for plant communities. Much earlier work was concerned with what were grossly polluting conditions causing rapid and obvious visual damage, while current studies are more likely to be focused on lower concentrations that may not result in visual damage but may affect yield, growth, or seedling success, or alter the balance of the plant community. Recent attention has been given to defining 'critical loads' of pollutant deposition as a strategy for control policies aimed at preserving some natural or exploitable resource.

As a start, it is important to distinguish the adverse effects of gaseous pollutants, such as sulphur dioxide, fluorine, or ozone, and those of acid or nitrate deposition. It has become popular to call all air pollution 'acid rain' (Rossi, 1984)—confusing and misleading since some damaging air pollutants are neither acidic nor present in rain. If pollution is to be controlled effectively and damage avoided or reduced to some acceptable level, the specific causal agents must be identified and their sources identified so that control or remedial action will be successful and economic. Effects may also be attributable to more than one cause, with possible additional or enhanced effects due to acid rain or other environmental stress agents; it is important to know to what degree the effects can be avoided by control of the causal agent(s), or by countering acid rain or other environmental conditions.

In most industrial countries, the present levels of air pollutants are dramatically lower than those of even a few decades ago (Lines, 1979), although some poorer countries still have little or no effective control of polluting emissions and may have conflicting economic aims. The strategy of diluting pollution by greater dispersion ('the solution to pollution is dilution') led, for instance in the UK after the 1952 smog, to larger but fewer sources discharging at greater height. This was notably successful in improving conditions in urban and industrial areas such as London and New York, hitherto grossly polluted (Brimblecombe, 1986). An inevitable corollary is that lower levels of pollution (and any effects) are distributed over a wider region. If this reduces concentrations below the effects threshold (see below), such a strategy is acceptable if there are no important long-term or persistent effects at lower exposure levels. The progress of pollution control in cities and by heavily polluting industries through Clean Air Act legislation and its manifest benefits to human health should be acknowledged—indeed the very success of this policy in Britain since 1959 has promoted some complacency, although public expectation is for still cleaner ambient air conditions with acceptable costs in the future.

Levels of some air pollutants, notably sulphur dioxide and chlorine, have been lowered in recent years by as much as a hundred-fold, but two pollutants, nitrogen oxides and ozone, have undoubtedly increased. The increase in emissions from vehicles, including aircraft, has been especially rapid, and there is good evidence that the frequency and intensity of high ozone episodes has increased by about 20–100% in the mid-latitude troposphere since about 1956 (Derwent et al., 1987). The effects of sufficient versus excess nitrogen gases or nitrate deposition have important consequences at both canopy and root level. In addition, carbon dioxide from fossil fuel exploitation and use, volatile organic carbon compounds from natural sources and solvent use, and ammonia from intensive farming are all part of the current pollution climate. While the damaging effects of sulphur dioxide have dominated concern hitherto, the present mix of pollutants becomes more significant with today's lower SO_2 concentrations. In the USA, recent assessment of air quality (Irving, 1991) shows that national emissions there of SO_2, NO_x and VOC are about equivalent. A secondary pollutant, ozone, also comes to the fore as being the most damaging to vegetation.

4.2 EFFECTS OF GASEOUS POLLUTANTS

The point has already been made that gaseous and aerosol pollution is not 'acid rain' and it is necessary here to distinguish their different effects, although this text concerns 'acid rain' and strictly need not include an assessment of gaseous pollutant damage. However, because current usage is so prevalent, and in the case of some reported pollution damage, viz. forest decline, since both pollutant gases and acid rain are implicated, it is appropriate to provide some account here of the damage to vegetation from gaseous pollutants. For most western countries with effective pollution control and commercial selection of tolerant varieties, crop damage is, to a large extent, 'yesterday's problem'. However, for the reasons explained above, the assessment of damage from air pollutants is useful here to establish its extent and its causes and mechanisms, and to identify effective control measures. Visible damage after sufficient exposure to a variety of pollutants is often not specific, so that information on ambient levels of pollutants is necessary.

Gross effects of high levels of exposure to gaseous pollutants have been obvious

Table 4.1—Mechanisms of air pollution damage and their timescales (after Taylor and Pitelka, 1992)

Stress	Mechanism	Timescale
Ozone, O_3	Community change	Days/weeks
	Stomatal closure	Minutes/hours
	Free radicals	Minutes to days
SO_2	Stomatal closure	Hours/weeks
	HS emissions	Minutes/hours
CO_2 excess	Community change	Days/weeks
	Stomatal closure	Hours/days
NO_x	Nitrate reductase	Hours/days
	Community change	Days/weeks

for several hundred years, even before the Industrial Revolution. It is a common observation that the degree of damage declines with distance from a source, and so with (implied or measured) concentration (Roberts *et al.*, 1983). More sensitive species, such as lichens or mosses, are often absent in conditions of high exposure, and are found with increasing frequency with distance from the source (Lee *et al.*, 1988; Looney and James, 1988; Armentano and Bennett, 1992). The changing pattern of pollution means that more subtle and interactive phytotoxic effects are important, particularly with regard to species or plant communities of conservation and scientific interest. Pollutant exposure will influence phenological adaptation and the genetic structure of sensitive species, with consequences on competition and biodiversity of the community (Taylor and Pitelka, 1992).

At current levels of pollution, investigation of effects is more complex than earlier studies of gross pollution damage. Many natural factors such as climate, soil type and cultivar sensitivity are involved, so that year-by-year differences and site variations have to be taken into account. Although experimental exposure techniques can now provide low concentrations and complex exposure regimes, it is still difficult to match field conditions with controlled exposures, while community studies are notoriously difficult to interpret (Armentano and Bennett, 1992). There is a place for a variety of approaches.

Mechanisms of damage

Mechanisms or processes by which plants respond to pollutant stress are an important aspect of understanding and controlling pollution so as to distinguish their single and combined effects and to evaluate the role of natural factors which modify response (see Table 4.1). The plant growth processes and the altered production and distribution of carbohydrates play a vital role in the plant's capacity for growth and its resistance to stress. Effects relate both to surface (cuticular) damage and to physiological/biochemical processes within the plant; secondary effects may also be seen at the root level.

Pollutant exposure at the leaf surface affects stomatal function—the stomata are

Table 4.2—Crop response to exposures to pollutant mixtures (after Irving, 1991)

Pollutant mixture	Test crop	Exposure conditions	Plant response
$O_3 + SO_2$	Beans	0.2 ppm each, 7 h d^{-1} over 4 d	Stomatal conductance, foliar damage, growth, photosynthesis
$O_3 + NO_2$	Beans, radish, wheat	0.08–0.1 ppm, NO_2 first, then O_3	Growth increase in radish, bean
$SO_2 + NO_2$	Beans	0.1 ppm each, 5 d	Transpiration increase with each; decrease with both
$SO_2 + NO_2$	Soybeans	0.13–0.42 ppm SO_2, 0.06–0.4 ppm NO_2	Chlorophyll reduction, yield reduced
O_3 + acid rain	Soybeans	Ambient O_3 + pH 2.8, 3.4, 4.0	Ozone reduces stimulatory effect of acid

the major point of reaction with the atmospheric environment. The nature of this response varies according to the pollutant, duration of exposure and environmental conditions such as temperature, light, humidity and wind, and stomatal opening or closing in response. Sulphur dioxide and NO_x alone increase transpiration, but decrease it in combination (Table 4.2). Sulphur dioxide dissolves in water to form sulphite ion (SO_3^-) and bisulphite ions (HSO_3^-), both highly reactive with disruptive effects on enzymes and structural proteins, and on nucleic acid replication. Nitrogen oxides penetrate the cuticle even when stomata are closed and can then react with the extracellular water to produce nitric acid, nitrite (NO_3^-) and free protons. Ozone causes stomatal closure but enters through the cuticle and cell membranes, generating secondary products such as hydroxyl radicals, and other oxidants. At high concentrations, necrosis, protoplast collapse and cell death ensue. At lower concentrations, accelerated senescence of foliage is seen, possibly involving enzyme synthesis, with reduced photosynthesis and increased dark respiration. Mechanisms of damage due to gaseous pollutants alone and in combination are reviewed by Wolfenden *et al.* (1992); detoxication mechanisms which counter the free radicals resulting from pollutant exposure have been identified.

4.3   EXPERIMENTAL INVESTIGATIONS

Experiments generally fall into one of three modes: fumigation in which the plants are exposed to a controlled source of a known pollutant; filtration in which the plant is exposed to monitored air quality, but with one or all pollutants removed; and transfer of a target species to a pollution gradient in the field. Each mode has advantages and disadvantages, and may provide somewhat different information. It is often best to combine information from a variety of studies and observations.

Fumigation procedures

Fumigation experiments subject a target species to the damaging agent in controlled conditions, possibly alone on a continuous or variable regime, or in combination with other factors, such as temperature, humidity or other stresses. The effects of sulphur dioxide, long recognized as a common and damaging pollutant, have been studied in this way with a wide variety of target species.

It is useful to consider exposure in terms of both the concentration of a pollutant and exposure duration (dose). **Concentration** is an important variable, since it is usually possible to determine a **threshold** concentration below which adverse effects are slight or absent, and high levels of concentration where effects are evidently acute or immediate (Fig. 4.1). Acceptable ambient concentrations can provide an environmental quality objective (EQO) for pollution control. However, **dose** over time may be a most important measure of chronic damage from low levels of exposure. Plants and their cultivars differ in their sensitivity to pollutant exposure, even between individuals of the same population, and response may vary with development from germination to maturity. Recognition that some damage may be reversible if exposure is abated, or when a sensitive stage has passed, is also important (Wellburn, 1988).

The design of an experimental protocol is crucial to obtaining meaningful results, especially to relate findings to realistic levels and patterns of exposure. It is important to focus on the effects of concern, whether these be, perhaps, acute foliar lesions, or reduction of growth or yield, specifying a reference (or threshold) standard in unpolluted conditions. Ambient air quality conditions prevailing during any experiment are needed, including daily and seasonal variation. Air quality standards are different in urban areas, where ornamental or garden plants are at risk, from those in rural areas where arable crops are cultivated or where species of conservation interest are present. The literature is replete with reports of responses of many plant species to air pollutants, but the effects of mixtures and of community responses at low levels of pollution found today are rather few.

Controlled fumigation experiments have familiar disadvantages—enclosed target species are subject to unnatural conditions, and infections by plant pests may invalidate the responses, stimulating the development of techniques for open-air fumigation (Roberts, 1987; Darrall and McLeod, 1990).

Filtration experiments

An alternative to fumigation exposures is to exclude a pollutant by charcoal filtration systems. Many of the early 'classic' experiments were made by comparing plant responses to removal of pollutants from urban air with those seen in the open, measuring growth or yield differences or the occurrence of visible symptoms. A combination of gases and particulates present in polluted urban air and their collective removal often showed that growth or yield improved by as much as 30%. Results were not always consistent with more controlled fumigation exposures, probably because of additional unidentified factors, with additive, synergistic or even beneficial effects. Another disadvantage of filtration experiments is that specific dose relationships for individual and/or combined pollutant mixtures are difficult to achieve.

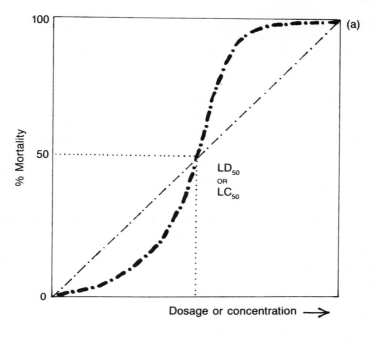

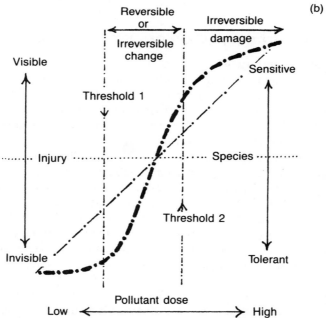

Fig. 4.1—Some basic responses to toxic agents (after Wellburn, 1988).

A further problem is that long-term exposures are subject to temporal variations including short-term transient peaks, for example for both SO_2 and ozone which damage plant tissues at the cost of growth or production.

Field surveys

Another approach is through field surveys, using both temporal and spatial gradients to explore plant or community response to ambient pollution climates. The value of this approach takes account of environmental factors and community interactions but is generally limited to the natural or unmanaged flora. For species of conservation interest, there is extensive documentation of a progressive decline or limited extent of occurrence of sensitive species. Thus, surveys of lichens in Britain over nearly 200 years have been associated with declining air quality, and wetland mires in western Britain are reported to be adversely affected. However, alternative explanations of change are not always explored, and interpretation of community responses also have to take account of natural variation and the adaptation potential of the species/varieties present (Armentano and Bennett, 1992). Comparison between surveys is also confounded by differences in methods and objectives. One of the most successful approaches is to select well-studied and maintained sites where sufficient information on wider ecological components is already available, and where observations can be repeated for a substantial period—a timescale of half a century is covered in the classical study of deciduous forest understory vegetation in Sweden (Falkengren-Grerup, 1986) and for forest trees in Ohio (Cribben and Scacchetti, 1977).

4.4 EFFECTS OF CURRENT AIR QUALITY ON CROPS AND TREES

Typical rural concentrations of sulphur dioxide in the UK (annual means) are now about 10 ppb (25 μg m^{-3}), with $<$ 5 ppb over much of the country, or even 1 ppb in areas remote from industry. In urban areas, concentrations are more variable, but typically 10–20 ppb (26–52 μg m^{-3}) (Irwin et al., 1990). Values for rural areas in the USA range from 2.5–10 ppb—the lower value is considered the minimum detectable and is characteristic of remote forested mountain areas (Irving, 1991). In only one urban area in the USA was the national standard of 30 ppb (annual) exceeded. For arable crop species exposed at 100 μg m^{-3} (about 35 ppb), grasses or cereals suffer some reduction in growth, while at lower levels both growth stimulation and reduction are reported (Fig. 4.2; Roberts, 1987). The effect of persistent long-term fumigation of a winter-growing grass at the highest rural levels in the UK (Table 4.3) was reduced growth, although this was overcome by later spring growth (Ashmore et al., 1988). In the USA, however, current ambient SO_2 concentrations alone are not responsible for any regional-scale crop yield reductions (Irving, 1991). In rural Germany SO_2 concentrations are 7–38 μg m^{-3}, but are higher in the Ruhr, 56 μg m^{-3} (Krahl-Urban et al., 1988); an intermittent exposure (300 μg m^{-3} for four hours per week over four growing seasons) of the natural ground flora of a German forest altered the community structure (Wolfenden et al., 1992).

Oxides of nitrogen in air include several species conveniently grouped as 'NO_x,' though this is somewhat misleading as the occurrence and effects of the different

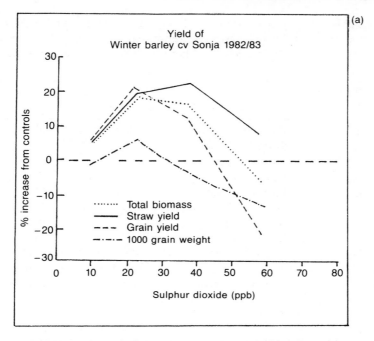

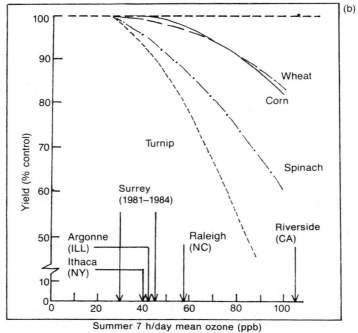

Fig. 4.2—Crop responses to two air pollutants: (a) SO_2 and (b) O_3 (after Roberts *et al.*, 1983; Roberts, 1987).

Table 4.3—Typical urban/rural concentrations of SO_2, NO_x and O_3 and their threshold concentrations (Roberts *et al.*, 1984; Darrall, 1989). U = urban, SB = suburban, R = rural concentrations, nd = non-detectable

Pollutant	Effect (expt.)	Threshold (ppb)	Air conc. (ppb)
SO_2	Photosynthesis	200–400	50 (U)
	Stomatal function	> 100	15 (SB)
	Assimilation	> 40	12 (R)
	Respiration	35–380	7 (R)
NO_2	Photosynthesis	500–700	78 (U)
	Stomatal function	< 100	18 (SB)
	Assimilation	—	8 (R)
	Respiration	35–380	nd (R)
O_3	Photosynthesis	100 in 1 h	15 (U)
		35 in 3 weeks	15 (SB)
	Stomatal function	< 200	20 (R)
	Assimilation	> 50	19 (R)
	Respiration	> 150 (stimulates)	

species are different. In experiments it is not always clear whether 'NO_x' exposure includes nitrous oxide (N_2O) as well as nitrogen dioxide (NO_2) and nitrogen oxide (NO). Nitrous oxide is the most abundant form, naturally produced by microbiological activity in soils in the 'denitrification' step of the N cycle. Nitric oxide is produced by high-temperature combustion by a combination of atmospheric nitrogen and oxygen rather than oxidation of the fuel nitrogen. In the atmosphere, nitric oxide is rapidly oxidized to nitrogen dioxide by reaction with ozone, reducing ambient ozone concentrations. Annual mean urban or industrial area concentrations of nitric oxide in the UK are now (1990) 16 ppb (31 $\mu g\ m^{-3}$) but levels may reach 40 ppb in London; remote area concentrations are only about 4 ppb (8 $\mu g\ m^{-3}$) (Irwin *et al.*, 1990). Exposure to NO_2 leads to negligible or statistically doubtful growth reductions at current rural or urban levels (Table 4.3). In the USA, maximum hourly values are 7 ppb, but 99% are < 5 ppb; NO_2 at ambient levels is not found to be a direct cause of regional growth or yield reductions in agricultural crops (Irving, 1991), and the national standard of 53 ppb (annual) was exceeded in 1985 in only one urban area. Levels reported in Germany are 50 $\mu g\ m^{-3}$ in the Ruhr and 8–21 $\mu g\ m^{-3}$ in forest areas (Krahl-Urban *et al.*, 1988) so forest damage is unlikely from NO_x exposure alone, although there is concern expressed that the ecosystem may be damaged by excess N supply (including NH_3) (Wolfenden *et al.*, 1992). This phenomenon is also thought to be the cause of changes in the balance of heathland species in the Netherlands, and in those of wetland mires.

The picture for ozone is rather different; ozone is generated in the atmosphere by reactions between oxidizing and reducing pollutants in the presence of humidity and sunlight. It has increased in recent decades and reaches peak concentrations in summer when stable air masses allow pollutant concentrations to build up over several days. These conditions occur during the growing season when ozone pollution has the greatest economic effect on commercial crops. A mean exposure at $\geqslant 50$ ppb reduces yield by as much as 10%—within the range of concentrations in southern England (Table 4.3), which exceed 60 ppb (120 μg m^{-3}) for 50–100 hours during the growing season (Irwin *et al.*, 1990). Growth reductions and some inhibition of flowering in native species is reported (Ashmore *et al.*, 1988). In the USA background levels of ozone are judged to be 20–35 ppb and even 'clean' sites may have 28–50 ppb, but many sites in the eastern USA exceeded 80 ppb (hourly mean) in 1988, and in California and the mid-West, hourly means were 300 and 200 ppb. Ozone is considered to be the cause of reduced agricultural production there (0–56%) and contributes to the decline of forests and to changes in coastal sage scrub in California (Armentano and Bennett, 1992). The hourly national standard (120 ppb) was exceeded in 110 counties (population about 145m) in 1985 (Irving, 1991). Concentrations in Germany are also high—23 μg m^{-3} in the Ruhr and 50–100 μg m^{-3} in rural areas (Krahl-Urban *et al.*, 1988). It is a feature of ozone distribution that concentrations are higher outside cities where NO_x generated by traffic provides a 'sink' for ozone.

Pollutants rarely occur in isolation, however, and the combination of sulphur, nitrogen and ozone gases may be more stressful than any alone (Table 4.2). Sulphur dioxide and NO_x have additive or synergistic effects, while for SO_2 and ozone effects vary from antagonistic to additive; it is not clear how important this is on a regional or national scale. In general, it is suggested from recent fumigation studies with mixtures at current concentrations, that responses could be greater than additive for a number of species or cultivars. More research is needed to evaluate ambient air conditions and to explain how synergism occurs. In addition to this possibility, the effects of climatic and biotic conditions (especially drought, frost and rate of growth) are thought to be important; in some cases stress from acid rain has been postulated (Table 4.2; Fig. 4.3).

4.5 EFFECTS OF ACID DEPOSITION (RAIN, MIST AND FOG)

While pollutant gases exert their effects at the foliar surface and enter plant tissues with effects at the molecular level, acid rain has direct physical and chemical effects at the leaf surface. The physical properties of deposition—as rain, snow or sleet, or as fog or mist—are important determinants of their transfer from air. Materials present as aerosols are characteristically less than 10 μm in diameter, while cloud and fog droplets are larger at about 60 μm; rain droplets may be several hundred or even a thousand μm. The smaller particles have low settling velocities and may remain suspended in the air for long periods, although possibly accreting with time; they are not readily captured by surfaces, whereas the larger mist or rain droplets deposit about 100 times faster. Climatic conditions, such as humidity, wind speed and rainfall pattern, the physical nature of the surfaces, and the complexity of the vegetation canopy all influence deposition, and so determine the intensity and

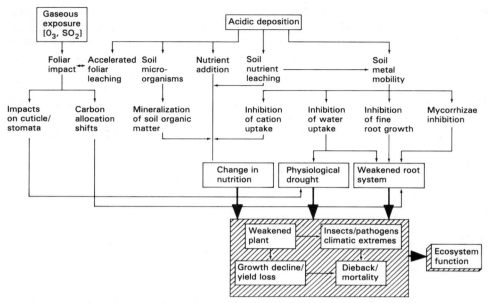

Fig. 4.3—Possible pathways and mechanisms of plant response to acid rain and gaseous pollutants (after Irving, 1991).

frequency, i.e. the dose, to which the target is exposed (Jacobson, 1984). In some circumstances, co-deposition of NH_3 and SO_2 also occurs (see Chapter 3).

The chemical constituents of rain are also important in determining effects, both beneficial and adverse. In addition to acidity (H^+), and acid anions such as sulphate and nitrate, ammonia and sea salts may also have significant foliar responses and may lead to altered growth and function, both in the canopy and at the root level. Sulphate is essential for plant growth and metabolism, and is not phytotoxic—indeed in sulphur-deficient areas, sulphur deposition may be an important contribution. Generally, atmospheric nitrate and ammonium also provide some of the essential nitrogen, and do not produce foliar lesions even at high dose, but absorption of ammonium in the canopy may induce more acid conditions at the root level (Miller, 1984). Foliar absorption of nutrient ions, including trace metals, is also an important route for plant uptake. This is generally of an intermittent nature, however, so continuous uptake via the root system provides the major supply (Fig. 4.4; Johnson *et al.*, 1982). In contrast, chloride, often not perceived as a pollutant, is considerably more toxic, as witnessed by 'salt burn' of vegetation in maritime areas or downwind of industrial sources.

Effects on crops and trees

Exposure of vegetation to sufficiently high concentrations of acidity (i.e. pH < 3.3, $[H^+] \geqslant 500 \ \mu eq \ l^{-1}$) can produce foliar lesions or leaf abscission in sensitive species. The surface characteristics of the foliage determine the rate of deposition, retention

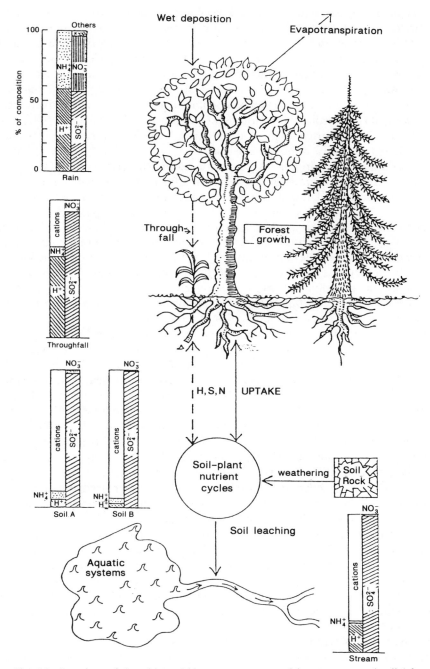

Fig. 4.4—Reactions of deposition within a tree canopy and between tree and soil (after McLaughlin, 1985).

and thus exposure, but rain following earlier aerosol deposition may wash off the accumulated deposit, so that the intensity and frequency of rain are important. High humidity increases aerosol deposition, exposure and the severity of the response. The size of the aerosol particle is also important, with smaller particles causing marginal and tip necrosis, and larger particles a more localized necrosis. Effects seen are chlorosis, wilting of the leaf tips, abscission of leaves and accelerated senescence (Jacobson, 1984). In continuous experimental exposures, growth reductions are evident with SO_2 aerosol concentrations of 10–100 μg m^{-3} (Roberts, 1984), about tenfold higher than reported levels of acidic aerosols and fogs in rural or urban areas (UK and USA) generally, although levels up to about 680 μg m^{-3} have occurred in heavy pollution incidents in Los Angeles and London (Roberts et al., 1983; Jacobson, 1984). Thus, although there is little concern for foliar damage with visible lesions, lower exposures may not reveal delayed responses to short-term variations.

A recent US review has demonstrated that, at levels experienced regionally throughout the eastern states, acid deposition is not responsible for reduction in crop yield. Further, there is no consistent association in time or space of damage with the incidence of brief phytotoxic concentrations in rain—only one report in the literature found visible injury as a result of three short-duration rain events (pH 3.8–3.9), even though hundreds of individual rain events have been monitored over more than a decade (Irving, 1991). In certain localized conditions, however, acidity is occasionally high enough to cause some visible injury and reduce the economic value of crops.

Plants naturally vary in their susceptibility; the needles of coniferous trees are less likely to show effects of exposure to acid mists than the leaves of deciduous trees or herbaceous plants. This is probably due to their waxy cuticle as well as their relatively small area. Conifer canopies are also exposed through the full year, and some foliage is retained so exposure is cumulative. Adverse effects (both foliar injury and reduced growth or yield) are found in herbaceous crop species at about pH 3.2–2.0 (beans, *Phaseolus vulgaris*), pH 5.6–3.0 (radish, *Raphanus sativus*) and pH 3.4–3.0 (a cereal, *Glycine max*) (Roberts, 1984), but possible acid effects on all tree species tested to date have been associated only with exposures at pH $\leqslant$ 3.0 (Irving, 1991). The range of response probably arises because of differences in the pattern of exposure, or from the use of cultivars of different sensitivity. Overall, however, visible damage seems unlikely at pH values greater than 3.0; lower pH is seldom reported except briefly as the first drops of rain fall after a prolonged dry period.

Foliage of annual plants is often rather resistant, and many plants can accept some reduction in photosynthetic area without evident restriction in growth. Reproductive systems or pollen germination (Cox, 1992) may be more vulnerable, especially in primitive plants where the reproductive organs are relatively unprotected (e.g. lichens and bryophytes). Unfortunately much of the research in this field has not distinguished gaseous pollutants from aerosols or acid rain, nor has it simulated realistic exposures for sufficient time.

The possibility that forest trees are subject to nutrient deficiencies indirectly by acid rain has often been raised, possibly through foliar uptake of potentially toxic metals, enhanced foliar leaching, or soil-mediated effects on root function. Tree ring analysis of conifer growth in areas subject to acid rain has yielded conflicting results:

in a Swedish forest survey, the growth of *Pinus sylvestris* and *Picea abies* is reported to be lower in areas of greater acid deposition and lower soil buffering capacity (Jonsson and Sundberg, 1972; Jonsson, 1977; Hallbacken and Tamm, 1986), but a similar study in Norway did not confirm this finding (Abrahamsen et al., 1976), nor did similar studies in the UK and northern USA (Cogbill, 1977). A more recent survey (1987–88) in the UK involving five tree species suggested that tree decline (crown thinning and yellowing) varies principally in relation to climatic conditions rather than to pollutants, including acid deposition (Innes and Boswell, 1987, 1989; Schulze and Freer-Smith, 1991)—indeed the effects of pollutants may be both beneficial and detrimental. In UK forests, crown density of Scots pine, oak and beech, but not Norway or Sitka spruce, is greatest in areas with heaviest acid deposition, and in northern Britain cyclical variations in spring droughts were related to year-to-year growth increments in *Pinus sylvestris* (Miller and Cooper, 1976). Thus, a direct causal effect of acid rain on forest growth is not established. Slow-growing conifers on poor soils are possibly affected by magnesium depletion of soils by acid rain, although growth in maritime areas is not affected since magnesium is supplied in sea salt aerosols. Acidic cloud water along with climate stresses affects some high-elevation spruce forests at the limits of their distribution in the eastern USA (Irving, 1991).

Acid rain or experimental acid application increases foliar leaching in a wide range of species, probably due to cuticular and cell damage, and enhanced H^+ and NH_4^+ exchange processes. The increased leaching is evident in greater ion concentrations below the canopy and in the lower nutrient status of the foliage. However, it is difficult to allocate the degree to which greater deposition and reduced root uptake from poor soils, rather than impaired canopy function, are responsible. In some cases where the nutrient level of the foliage is unchanged, it is proposed that the only effect is to increase the rate of ion turnover (Miller, 1984). Resolution of these inconsistencies might be expected from observations being made at experimental sites where exclusion of acid rain, or fumigation with pollutants including acid mist, has been studied in recent years.

Reduced foliar and root nutrition are thought to be another effect, although in nutrient-poor soils the additional nitrate in acid rain stimulates growth in both laboratory tests and in nature. The toxic effects of aluminium mobilized in acid soils (Ulrich, 1983b) was earlier invoked as a cause of declining tree health through root damage, but this now seems unlikely unless soil concentrations reach as high as $20 \, mg \, l^{-1}$ (Abrahamsen, 1984); whether the aluminium is present in toxic forms is not certain. Long-term changes in soil chemistry are expected from the cumulative effects of acid rain (see Chapter 5), but a wide variety of response in soils occurs and soil changes may be rapid and not necessarily matched to acid loading (Johnson et al., 1991).

Damage to the mycorrhizal status of forest trees as a result of acid rain has been claimed (Ulrich, 1983b) on the basis of samples taken in areas of forest decline. In an experiment where below-ground biomass (roots and mycorrhizae) was influenced by acid conditions, with an 'acid loading' ($1500 \, mm \, yr^{-1}$ at pH 3.0) about 30 times greater than that received at the source site, Scots pine showed some difference in root length and numbers of root tips (Dighton and Skeffington, 1987). Mycorrhizal types were altered to a small degree and this may have influenced mineral and nutrient uptake by the trees, and the Ca/Al ratio in soil solution will have changed in the more acid conditions, but the above-ground biomass was not significantly different between control and acid conditions. However, changes in root penetration with depth may lead to their more superficial distribution, with greater sensitivity to drought and storm damage. These consequences are often interrelated and the significance of the observations over the large spatial and temporal scales required for forest surveys is not at all clear.

Exposure to acid rain and ozone may be synergistic or additive. As with arable crops, ozone is a powerful pollutant with direct action on foliage, sometimes leading to visible injury, although the leaf tissues may respond to 'quench' the effects of the free radicals produced. This is a significant cause of forest damage in areas in the USA and where ambient levels in the growing season are $\geqslant$ 100 ppb, and where sulphur dioxide and NO_x at current ambient levels are not damaging. Sensitivity to ozone is greatest in magnesium-deficient areas, possibly an effect of acid rain on soils, but ozone also promotes magnesium leaching from canopies (Krahl-Urban *et al.*, 1988). However, exposure to SO_2, ozone and acidic mist has still failed to demonstrate long-term effects at current levels (Schulze and Freer-Smith, 1991).

In conclusion, although a few sensitive arable crop species may show necrotic lesions with acid exposure at pH 3.0, alone or with other pollutant exposures, in the range of acid rain pH from 5.0 to 3.0 this seems to have little effect on whole plant growth or metabolism. Only ozone is considered to be the significant agent of both crop and forest damage at levels found at present, possibly enhanced by acid rain.

Lichens, mosses and other natural plant communities

Lichens have long been used as indicators of 'air pollution'—a variety of acid gases as well as fluoride and acidity might provide a 'pollution climate' near to cities and industry. Species found at a variety of sites distributed around Britain were matched by Hawkesworth and Rose (1970) to levels of SO_2 in ambient air. Although potential effects of NO_x and ozone were acknowledged, national data for these pollutants then were few and no correlation was found (Looney and James, 1988). However, distribution of some species (Fig. 4.5) might match rainfall (see Fig. 2.3), even though high rainfall sites are those with highest acid loading. The return of some less sensitive species to urban environments is recorded as city air improved with the implementation in the UK of the Clean Air Acts (1956 and 1968), while continued decline in species such as *Lobarion pulmonaria* and *Parmelia* spp., and absence of *L. scrobicularia*, with a predominantly westerly distribution, have been attributed to acid rain, and are perhaps consistent with the current level of acidity in rain. The response of lichens to ozone has seldom been investigated, but tolerant species appear unaffected even at 7 mg m^{-3} and sensitive ones are damaged at about > 1 mg m^{-3} (about 500 ppb) (Looney and James, 1988).

The relative paucity of mosses, liverworts, *Sphagnum* and other moorland vegetation in British uplands, especially the Pennines, has been attributed to gross pollution levels in the past, as well as to continued unfavourable conditions. A hundred years

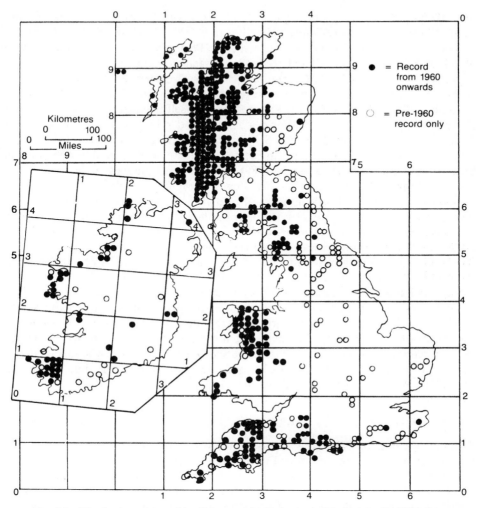

Fig. 4.5—Distribution of a sensitive lichen species (*Lobaria scrobiculata*) in the UK before
1960 and in 1980 (after Looney and James, 1988).

ago, SO_2 concentrations in the southern Pennines were reported in the nearby urban
environment (Manchester) to be in a range rising to 10 mg m^{-3} (3.7 ppm), implying
that even in rural locations there, concentrations could be at damaging levels. These
conditions prevailed at least until the mid-1950s (Lines, 1979). Most upland habitats
are characterized by climatic stresses and a high incidence of acid fog and mist, and
in Britain are often overused for recreation, adding to stress.

The flora of ombrotrophic bogs, which receive only rain as a source of water and
nutrients, is limited by excessive acidity and reduced calcium, magnesium and
potassium retention (Johnson *et al.*, 1991). Exchange of metal cations for H$^+$ in

growing *Sphagnum* and rain at pH 4.0–4.2 are about equally important (Clymo, 1984). Since peat is always more acid than rain (pH 3.35–2.8) an adverse effect of additional atmospheric input is unconvincing. Peat pH and base saturation were found to be inversely correlated to modelled acid loading at a range of sites in Scotland (Skiba *et al.*, 1989). Community changes also seem unlikely to result from acid rain *per se*, since mire ecosystems support acid-generating *Sphagnum* spp. (Bache, 1992). Paradoxically, it is claimed that *Sphagnum* itself is sensitive to acid rain, even though it lowers the pH of soil/bog water; the 'acid rain' response is really due to conversion of gaseous or deposited SO_2 to sulphite, SO_3^-, a powerful and reactive reducing agent. Vegetation changes in affected areas can be mimicked by experimental SO_2 or acid exposure (Lee *et al.*, 1988). Nonetheless, changes in plant life have certainly been observed and alternative causes might be explored. The failure of *Sphagnum* recovery in the Pennines is also attributed to increasing NO_x pollution, the moss taking up NO_x directly through the leaves (Lee *et al.*, 1988). This ability may become a disadvantage in areas affected by pollution since it depends on the induction of the enzyme nitrate reductase—successive additions of nitrate cause a fall in the cumulative nitrate reductase activity and accumulation of ammonium in the tissues (Wolfenden *et al.*, 1992).

Repeated studies of forest understorey vegetation undertaken in an area subject to acid rain in southern Sweden show changes in species composition, with species richness increased due to the invasion of nitrophilous species associated with higher soil nitrogen, probably due to increased atmospheric input (Falkengren-Grerup, 1986). Community changes in coastal sage scrub are reported in California, due to both high levels of ozone in summer, and increased SO_2 levels, but where acid deposition is low (Armentano and Bennett, 1992).

4.6   FOREST DECLINE—'*WALDSCHADEN*'

Although the possible effects of gaseous pollutants at current ambient rural concentrations on crops now seem trivial in Europe and North America, in public perception a serious problem in many countries is the loss or decline of forests (*Waldschaden* or *Waldsterben*), often attributed loosely to 'acid rain'. In 1980 and 1981 scientists and the media alerted public attention to conditions in West German forests, which were seen to be affected by a novel and mysterious disease which yellowed the foliage, caused its premature fall, and resulted in thin or abnormal and ugly 'stork's nest' crowns. This was not an entirely new phenomenon, since the symptoms had been seen in mature stands of silver fir (*Abies alba*) for a decade or more. At the earlier time the incidence of these symptoms was thought to be localized and was attributed to insect attack, or local pollution sources. In 1981, however, the disease was found to occur in Norway spruce (*Picea abies*) over a wide geographical area and in a variety of other species. Pessimistic predictions were that all of Germany's forests would soon be destroyed with progressive and possibly irreversible symptoms. Both generalized air pollution and acid rain were blamed, possibly exacerbated by extreme climate events, but many other potential causal agents were also put forward to explain the dieback (Blank, 1985; Schulze and Freer-Smith, 1991).

Public debate became emotional and demanding on this matter, fuelled by political initiatives which sought to exploit the strong national feeling for forests in the German people. The ensuing clamour perhaps concealed the almost total lack of reliable information about the true symptoms, its rate and spread, and its possible causes. Damage was being reported in other European countries and North America by 1982, adding to the growing demands for 'action'. The apparent spread of the disease, and widening public concern, led to the initiation of substantial national and international research programmes with the aim of improving information and understanding.

Extensive surveys were made in some areas of Germany in 1982 and 1983, followed by a more standardized annual survey in 1984 set up by a national authority. Differences in survey design and methodology meant that the results of the first two years were not comparable with those that followed, nor with those employed elsewhere, confounding any trend analysis (Blank, 1985). Improvements in observation and quantification of damage allowed its extent and severity to be established more objectively. In 1984, about half the forest area in western Germany was classed as 'damaged', about a third as 'slightly damaged', and about a fifth as suffering 'medium–severe damage' (Fig. 4.6). These classes were defined from the estimated loss of foliage:

Class	Status	Needle loss
0	Healthy	< 10%
1	Slight damage	10–25%
2	Moderate damage	26–50%
3	Severe damage	> 60%
4	Dead	100%

Standard plots were selected so that annual records would follow the progress of the disease across the nation. Since 1984, damage classes were revised with class 1 considered only an 'early warning stage' to account for natural variability in tree crown condition and the year-to-year response to climatic conditions. As a result, the cumulative area considered as damaged was reduced from 50% to 20% (Blank *et al.*, 1988). Another improvement was a more objective approach—initial surveys had requested attribution only to damage caused by air pollution, implying that such a causal link was already established to the exclusion of others, but many reports came from areas with scarcely acid rain and low levels of SO_2 and NO_x, such as the Black Forest.

As similar damage was reported for other countries in Europe, forest surveys were initiated by the UNECE Convention on Long-Range Transboundary Air Pollution, starting in 1986, following a joint US EPA and German joint assessment in 1984. The findings confirmed those of national surveys, although national differences in methodology, in tree species and communities, as well as in soils and climate preclude generalizations. Only some symptoms—yellowing, needle loss and radial growth reduction—are found in common, and are general rather than specific to any cause. The overall picture for western Europe is that between 1% (Italy) and 25% (Netherlands) of forests show > 25% defoliation. In the UK > 40% defoliation was

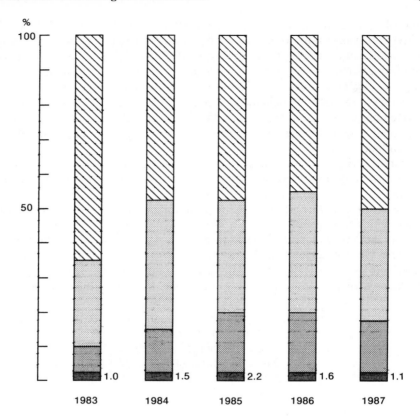

Fig. 4.6—Estimated forest damage in western Germany (FRG), 1983 to 1987 (after Blank *et al.*, 1988).

seen in 1987 in Sitka spruce, Scots pine and Norway spruce, and deciduous trees such as beech and oak showed signs of crown thinning (57% and 66% respectively) (Innes and Boswell, 1987, 1988); it was argued that crown density loss up to 25% or more occurs naturally in the UK, and does not damage the tree significantly. Indeed, a variety of studies show that there is a general *increase* in crown density of Scots pine, oak and beech in areas with *higher* pollution. Although counter to the European acceptance of causal link with air pollution, this observation is consistent with the many reports that growth in excess of predictions was observed with low levels of exposure to sulphate, ammonia and nitrogen oxides.

Initial predictions of a progressively severe and wider spread of forest damage in Europe have not been fulfilled; indeed, since 1985, there is evidence of recovery (Huettl, 1989), with volume increments in spruce stands 20–40% higher than predicted (Kenk and Fischer, 1988). The majority of forests in North America are not in decline (Irving, 1991) and there is no evidence that acid deposition or associated pollutants alter the resistance of red spruce to winter injury. Sugar maples in the USA and Canada are also mostly unaffected and involvement of acid rain and/or ozone has not been demonstrated. A decline of southern pines has also not been shown— reductions in tree growth are reasonably attributed to land-use patterns, and its magnitude is not related to the degree of acid deposition. In California, ozone is the key factor in decline of pines. Altogether, acid rain is now considered in the USA as a minor factor affecting health and productivity of forests, with the exception of red spruce at one high-altitude site.

Experimental studies have tested the effects of climatic and air pollution regimes in both controlled and open-air exposures. A joint study in Germany, focusing on Norway spruce grown in environmental chambers (Blank *et al.*, 1990) and exposed to ozone and acid mist (pH 3), found that these conditions could not explain forest decline, although some metabolic changes were observed. A wide variety of response could be attributed to clone sensitivity, soil character and needle age, rather than to pollutant exposure. Schulze and Freer-Smith (1991) review the possible causes of forest decline in Europe, but are cautious about statistical deductions on a stand level; they note that large pollutant concentrations often have short-term rather than long-lasting effects, but that S and N deposition modifies soil chemistry and plant nutrition. Open-air fumigation at realistic exposures may help to integrate the effects of these factors (Darrall and McLeod, 1990).

Various hypotheses have been proposed to explain forest damage (Table 4.4). It is generally accepted that **air pollutants** (SO_2, HF, NO_x and O_3) separately or in combination cause changes in stomatal behaviour, in turn affecting water balance, altered C and N assimilation, root growth, and winter sensitivity, especially in conditions of environmental stress (Wellburn and Darrall, 1990; Wolfenden *et al.*, 1992). The complexity of interactions (Fig. 4.5) makes attribution of cause, or prediction of the direction or scale of change, almost impossible. However, significant advances have been made in dissecting out the aetiologies of various kinds of forest changes—for Norway spruce, five damage types can be distinguished (Table 4.5). These can be reasonably related to a limited variety of causes: type 1 (chlorosis) with magnesium deficiency, type 2 (crown thinning) with climate, type 3 (needle reddening) with fungal infection, type 4 (chlorosis + needle loss) with potassium and manganese deficiency, and type 5 (crown thinning) with coastal/maritime conditions.

Type 1 (chlorosis) seems to be the best established and has been recorded for many sites world-wide, although not in the UK or Norway (Roberts *et al.*, 1989) possibly because of an abundant maritime Mg supply. **Magnesium deficiency** leads to foliar loss of magnesium, impairing photosynthesis and carbohydrate translocation, leading to greater sensitivity to fungal pathogens, drought and cold stress. It is associated with low soil magnesium, an effect remedied by the application of magnesium-containing

Table 4.4—Hypotheses to explain forest decline

Possible cause	Symptoms	Effects
Acid rain/soil acidification	Low soil cations	Nutrient deficiency
		Foliar leaching
	High soil Al	Weakened root system
	Acid pulses	Toxic metal releases
		Growth decline, dieback
Airborne gases (O_3, SO_2)	Foliar damage	Foliar lesions
	Water stress	Stomatal dysfunction
	Photosensitivity	Chlorophyll breakdown
		Slow growth
Excess N deposit	Growth stimulation	Sappy foliage
		Nutrient imbalance
	Low frost/drought resistance	Water stress
		Frost/drought damage
	Low pest resistance	Pest damage

Table 4.5—Damage types for Norway spruce (*Picea abies* L., Karst) (after Blank *et al.*, 1988)

Type	Description	Symptoms	Occurrence
1	Yellow needles at high altitude	Chlorosis, loss of needles, Mg deficiency	Mountains in central/south Germany, Black Forest, Bavaria
2	Thin crowns at medium altitude	Needle loss, little chlorosis	Central/north Germany, 400–600 m
3	Red needles, older stands	Yellow to red needles, then shedding	Foothills of Bavarian Alps
4	Chlorosis of needles	Chlorosis and needle loss; K or Mn deficient	Calcareous Alps about 1000 m
5	Crown thinning	Reduced growth, needle loss	North German Plain, coastal areas

fertilizers or lime (Huettl, 1989).

Multiple stress, **excess nitrogen** and **soil acidity** have all been proposed to explain other damage types. Multiple stress has been invoked as a cause of impaired plant metabolism and increased sensitivity to other factors. Soil acidity coupled with high aluminium in soil was an early hypothesis exemplified by conditions seen at Solling Forest in Germany (Ulrich, 1983a), where high soil aluminium was associated with root damage. However, this does not explain regional decline where soils are neither acid nor high in aluminium. Further, conifers and other tree species have a high threshold for aluminium (Abrahamsen, 1984), although a critical Ca:Al ratio may influence root distribution which may have profound effects on water balance as well as on nutrient uptake (Schulze and Freer-Smith, 1991).

A third causative agent is **ozone**. There is good evidence for increased ozone concentrations in forest areas of Europe and North America, in association with higher NO_x and hydrocarbons. However, the laboratory studies (above) fail to simulate the symptoms of forest decline and support for this explanation is limited to sites in California.

A final hypothesis is that *excess nitrogen deposition*, either gaseous or in acid rain, increases the susceptibility to climatic and biological stresses and accelerates soil leaching (Agren and Bosatta, 1988). This is substantiated for the Netherlands where N deposition may be in excess of $50\,kg\,N\,yr^{-1}$, mainly from intensive animal husbandry. This relates to type 5 damage in spruce but is not so clearly implicated in other types of damage, especially at high altitude (Blank *et al.*, 1988).

Development of this more analytical and objective approach in the past few years has clarified this issue, including the possible ancillary effect of acid rain. Forests in central Europe are damaged significantly but silvicultural and management practices are available to counter damages due to nutrient disorders. Forest decline is now less evident than in the early 1980s perhaps because of better management and/or changes in climate; demands for action are less strident. But if the condition of forests declines in the future, it is still not certain that we will have sufficient knowledge to overcome it without further investigation.

4.7 UNRESOLVED ISSUES

Confusion beween 'air pollution' and 'acid rain' is misleading and hampers the identification of specific causal agents and necessary remedial measures.

The effects of gaseous air pollutants and acid rain on sensitive upland sites are not entirely resolved since acid conditions result from acid rain *and* acid production *in situ*. Transfer of material between sites is a useful technique but does not take account of clonal tolerances and differing microclimates. Comparisons between studies are often invalidated due to different methods and objectives.

The perceived effects of acid rain on lichen distribution suggest that other factors may be involved, but their relationships to lichen distribution have not yet been tested.

The contribution of air pollutants and/or acid rain to forest decline is not clear aside from type 1 (magnesium deficiency) effects where soil leaching may have occurred. Decline is not always closely associated with high levels of air pollution or acid rain.

Acid rain at some sites may lead to progressive soil acidification with effects on trees, but processes and their significance have not yet been tested critically on a stand or forest level.

4.8   SUMMARY AND CONCLUSIONS

The effects of gaseous air pollutants on crop species are now well established and quantified, either alone or in mixtures. At current ambient concentrations in rural areas of Europe and North America there is no reduction in growth or yield—indeed low concentrations have led to some increase.

In urban conditions, ozone may affect sensitive species, even though ozone levels may be moderated there by high NO_x emissions. In more distant areas where NO_x is lower, ozone can build up to damaging peaks during the summer growing season, with effects on crops and trees.

At observed levels of acidity in rain, there are no significant effects on vegetation. In Europe, acidification of soil is implicated in forest decline, possibly through enhanced mineral leaching. In the USA, however, acid rain has not been associated with forest decline except at one site.

Postulated acid rain effects on upland moors and wetlands are attributable to surface reactions with gaseous pollutants, to sensitive and localized clones, and possibly to higher ozone levels.

Forest decline is not a single disease with a specific cause. Climatic extremes have been invoked to explain the coincidence of decline at many sites through Europe. It now seems to have stabilized, and even to have been reversed at some sites. Evidence for forest decline in North America seems slight and not associated with acid rain, but forests are affected by ozone in California.

Forest response to magnesium deficiency, possibly enhanced by soil acidification, also reflects intensive forestry practice and can be reversed by appropriate fertilization regimes.

The additional nitrogen in polluted air enhances growth of forest trees and to some extent counters the possible growth check due to SO_2 exposure. No significant effects of O_3 have been seen at urban levels, and its effects are not critical in rural areas.

5

Acid rain, acid soils

5.1 INTRODUCTION

The importance of soils as the route by which atmospheric deposition and throughfall reach surface waters cannot be overemphasized. This was the substance of a strongly defended claim by a Norwegian geologist, I. Th. Rosenqvist, who published his book *Acid Soil/Acid Water* in 1977. He drew attention to the magnitude of acid-generating processes in soil which were greater than the annual input in rain (Fig. 5.1) and to the importance of rapid ion-exchange mechanisms in soils, influencing the soil solution (Rosenqvist, 1978, for summary in English), a claim challenging the view then current that water acidity and sulphate were directly related to rain composition and not influenced significantly by soil characteristics. The arguments that followed led to a reshaping of national research programmes so as to give greater emphasis to the role of soils. Fifteen years later, this is accepted universally and many acidification models and the current EC 'critical loads' strategy for acid rain control (see Chapter 11) are based on the different sensitivities for acidification of soils (e.g. Hornung, 1983; Hornung *et al.*, 1990a; Langan and Harriman, 1993). Soils are the source of minerals solubilized in the soil solution and of organic materials from decayed biota, both important to the quality of surface water and for its effects on aquatic organisms. It is now clear also that many natural soils can be highly acidic owing to the action of vegetative growth and bacterial processes (Fig. 5.2, Stumm and Morgan, 1981; Krug and Frink, 1983); for instance, podzol or peaty soils in Scotland are as much as one pH unit lower than that of the incoming deposition (Cresser and Edwards, 1987).

At this point it should be explained that soil acidity is difficult to define and measure. It can only be measured indirectly by measuring the pH of a soil paste made up in water or an electrolyte (either in water or a $CaCl_2$ solution)—the latter gives a lower value. In comparing data from different sites and times it is important to know what method has been used. Soil acidification may perhaps be better defined chemically in relation to the loss of base saturation (%BS) or as an increased 'lime requirement', or a decreased cation exchange capacity (CEC), as described further below.

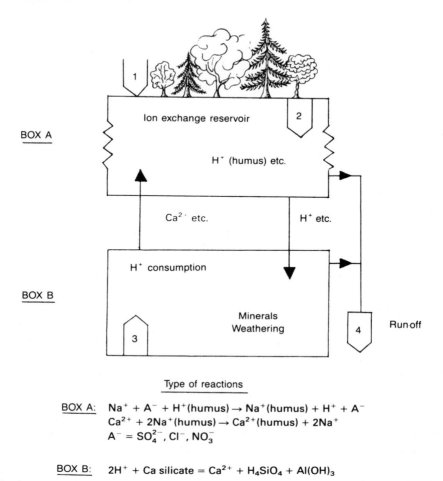

Type of reactions

BOX A: $Na^+ + A^- + H^+(humus) \rightarrow Na^+(humus) + H^+ + A^-$
         $Ca^{2+} + 2Na^+(humus) \rightarrow Ca^{2+}(humus) + 2Na^+$
         $A^- = SO_4^{2-}, Cl^-, NO_3^-$

BOX B: $2H^+ + Ca\ silicate = Ca^{2+} + H_4SiO_4 + Al(OH)_3$

Fig. 5.1—Acid pools and transfers through soils (after Rosenqvist, 1978). A⁻ represents strong acid anions present; 1 is the input from precipitation, 2 from bioproduction, 3 from weathering, and 4 loss to runoff.

5.2   THE NATURE OF SOILS

Soils are principally the product of the breakdown by **weathering** of geological materials, either the bedrock or the drift materials brought in from a more distant location. Other constituents come from the decay of plant and animal materials in the upper soil horizons. The idea that soils reach an inactive, steady-state or inert condition is inconsistent with changes observed over time—in fact they are quite dynamic, affected by climate and physical or chemical processes, as well as by seasonal vegetative growth, and are the 'home' of living roots, important microbial processes, and a wide variety of invertebrates and burrowing mammals. The rate at which soils are changing calls for re-evaluation (Johnson *et al.*, 1991), not only in areas subject

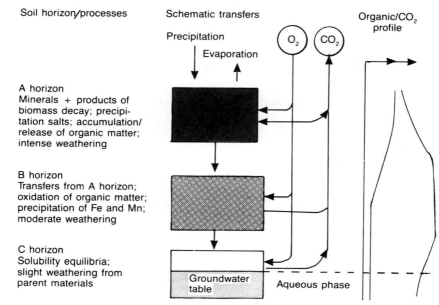

Fig. 5.2—Soil horizons and profile of organic materials, and carbon dioxide (after Stumm and Morgan, 1981).

to acid rain; even in pristine areas natural leaching and vegetative growth can create highly acid soil environments.

Because of the diversity of their origin and the conditions prevailing within the soil, soils are highly heterogeneous even within a small area—the relief or altitude of the site, its slope and aspect, soil porosity, as well as the amount and chemistry of rainfall all have an influence. This poses particular problems for sampling even qualitatively, and for prediction of catchment/site response to changes in acid rain. Soils are also important reservoirs for deposited sulphur and nitrogen, probably slowing the expected response to pollution control.

Biological activity in the soil plays an important role along with physical (and hydrological) and chemical processes. Over time, undisturbed soils become progressively more acid as a result of these processes and develop a strong depth gradient, with more organic humus materials near the surface, through to mineral horizons, down to the unweathered material of the parent bedrock (Johnston *et al.*, 1986; Fig. 5.3). The transition layers, or **depth horizons**, have different chemical characteristics and responses to atmospheric input or the throughflow of soil water. The chemistry of the soil solution reflects these characteristics and runoff its route and flow.

5.3 WEATHERING AND CATION EXCHANGE

The upper and lower horizons of soil are intimately related through hydrological routes, transferring the products of mineral weathering upwards, and of organic

(a) Podzol

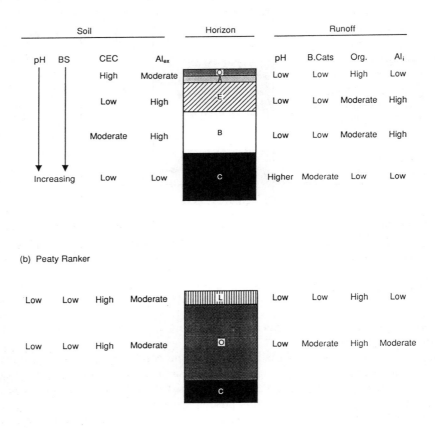

Key: O - organic layer, A - humified organic matter,
     E - eluvial layer, B - subsurface or illuvial layer,
     C - parent material, L - litter,
     B.Cats. - basic cations, Org. - organic constituents,
     Al_{ex} - exchangeable aluminium, Al_i - inorganic aluminium,
     BS - base saturation, CEC - cation exchange capacity

Fig. 5.3—Profiles of pH, exchangeable cations and aluminium in a podzol and a peaty ranker soil (after Skeffington, 1992).

activity downwards. In some conditions, however, where there is waterlogging of deeper soils, or formation of an impervious layer (podzolization) such as an 'iron pan', there is limited transfer between horizons. Silicate **mineral weathering** of primary bedrock materials comes about by physical processes, such as glacial or volcanic action, by hydrological processes exposing new materials, and by frost and drought, creating progressively finer, hydrolysable materials producing dissolved cationic species. These can be displaced by hydrogen ions generated by carbonic acid from rain or biological activity or organic acids, principally formed in the upper horizons; mineral acids in acid rain also contribute to the weathering process. The release of base cations is neutralizing, i.e. they consume the hydrogen ions—the lower the pH of the soil solution, the greater the release of cations. Acidity is thus higher in the humic surface layer and upper organic horizons of soil; weathering in the deeper horizons supplies mineral materials from drift or bedrock to neutralize it. This leads to a characteristic gradient through a soil profile (Fig. 5.3); organic material and pH decrease with depth and, in the absence of biological or chemical fixation, weathered cations are redistributed between the soil exchange sites and the soil solution and pass eventually to runoff. Soils with a sufficient supply of base cations from weathering can absorb and neutralize the acidity generated within the soils and from acid rain without acidification, but where acid inputs are in excess of this capacity, the soil becomes increasingly acidic. The natural accumulation of biomass in vegetation and humus, as well as progressive leaching of soluble materials from upper soil layers, is also acidifying.

Weathering rates vary according to the nature of the bedrock and their overlying soils. In Scandinavia, estimates for sensitive areas are 5–15 meq m^{-2} yr^{-1} (Jacks, 1990); on silty–sandy tills the rates are 20–50 meq m^{-2} yr^{-1}, and for soils derived from granites and gneiss, 20–30 meq m^{-2} yr^{-1}. Weathering rates are proportional to the area of mineral exposed, temperature and soil moisture, as well as pH (Warfvinge and Sverdrup, 1992).

Mineral weathering of many soils leads to the formation of clay minerals, principally aluminium silicates. These have the important property of adsorbing cations on their negatively charged surfaces; similarly products of biological decay, such as humic acids, bind with base cations. Together they provide a capacity for retaining cations— the **cation exchange capacity** (CEC). Most of this 'capacity' is taken up by the common base cations (Ca^{2+}, Mg^{2+}, Na^+ and K^+) which contribute to the **base saturation** (BS) of the soil. Hydrogen and aluminium in the soil constitute an **exchangeable acidity** (Bache, 1982, 1984); in extremely acid soils ($< 10\%$ BS), Al^{3+} and H^+ dominate and leach to drainage waters while in less acid soils ($> 20\%$ BS) base cations dominate and drainage waters maintain their acid neutralizing capacity. With progressive loss of base cations from the exchange complex (if not replenished by weathering), even less acid soils will become more acid with time. Soils of high base saturation are more fertile than those of low base saturation, explaining the benefit gained by farmers by liming as an agricultural improvement.

The slow weathering process and transfer of soluble soil constituents between horizons can be contrasted to the rapid cation exchange system; this means that following storms, or wetting-up after drought, there can be a rapid depletion of the soil exchanger with removal of released cations to the soil solution and to drainage. This is followed by the slower build-up of ions to the soil particles. In areas of heavy rainfall, such as parts of the western UK, and where bedrock is poor in basic minerals and soils thin, the potential for cation recharge of the soil exchanger is low. These are the areas considered most 'at risk' from acidification; sensitive areas of the UK and other European countries have been identified. In most areas, however, soils contain a reserve of weatherable minerals sufficient to provide that needed for plant growth and for neutralization of acidity.

Cation exchange of base cations at the surface of the clay particles and soil organic matter is crucial to the composition of drainage water. The cations on the soil exchanger control the composition of the soil solution in the short term but it is the anions present that provide their mobility. *En route*, some of the cations (including NH_4^+) are captured by root activity; if the uptake of cations exceeds that of anions equivalent amounts of H^+ and Al^{3+} are released (Fig. 5.3). The consequences are evident in soils below forests, especially at sites where weathering is slow.

5.4 CRITICAL LOADS

The critical nature of soil acidification and its determinant effect on surface water quality has been followed up with the adoption of the '**critical load**' approach to emission control. A threshold for acid rain effects emerged in 1982 when earlier scientific studies were reviewed at the Stockholm Conference (SNV, 1983); it was claimed on the basis of data from European and North American sites that in sensitive areas, waters of alkalinity $< 50 \mu eq \, l^{-1}$ will not be acidified if they receive a sulphur deposition of $\leqslant 0.5 \, g \, S \, m^{-2} \, yr^{-1}$. This threshold was later called the 'critical load', and a dose–response relationship was implied. The critical value is exceeded when harmful effects (in this case, loss of alkalinity) on a receptor are identified.

Application of this concept (Bull, 1991) has proved difficult since 'harmful' is difficult to quantify except in terms of a specific target (e.g. to protect a community, a species, or a life stage); loading also implies some averaging in space and time and may not be applicable to a specific site or to short-term (e.g. acid pulse) effects. Further, damage often results from mixtures of pollutants and natural conditions such as drought or frost. It follows that a defined critical load has to be matched to a specific objective and conditions.

The critical load was first defined by Nilsson (1986) as 'the highest load that will not cause chemical changes leading to long-term harmful effects on the most sensitive ecological ecosystems'. He argued that the acid input should not exceed the alkalinity (i.e. weathering rate) of the system. Thus for soils, attention is focused on the input of H^+ and weathered alkalinity with the release of H^+ and Al^{3+}, along with mobile anions from the soils. A different approach was adopted with regard to N deposition to forests, with nitrogen saturation (in excess of growth demand) and subsequent leakage to runoff; the critical load in this case is that at which forest production is not increased.

Efforts have been made to define important dose thresholds. While there are ample data for sulphur deposition (which has little biological impact), this cannot be extrapolated directly to H^+ deposition (which does have impact), nor do we have sufficient understanding to define a critical load for nitrogen (Skeffington and Wilson, 1988). The ability of soil to use or retain N depends on the relative sizes of sources (atmospheric N, N mineralization) and sinks (plant uptake of N and denitrification), both influenced by climate and season. Biological processes involving N fluxes vary considerably and are influenced by season, previous disturbance, catchment management and vegetation. The ability to identify catchments or regions which are likely to become nitrogen-saturated is limited by lack of quantitative information about both the processes and the catchment characteristics that influence them. Nonetheless, sensitive areas (identified on the basis of rain input and soil data) have been mapped for European countries and this has been extended by identifying sensitive sites where critical loads are exceeded for soils and surface waters (e.g. for the UK, Fig. 5.4; CLAG, 1991, 1993); MAGIC modelling (see Chapter 11) is used with soil sensitivity values to derive critical load values for $10 \times 10 \, km^2$ areas throughout the whole country. This spatial resolution, however, will not match the scale of heterogeneity of soils in many areas. An alternative approach assuming steady-state soil chemistry (PROFILE) has been used to calculate the critical load to maintain soil and water conditions at Gardsjon, Sweden (Warfvinge and Sverdrup, 1992). The critical load approach cannot yet accommodate conditions arising from short-term episodes.

5.5   THE INFLUENCE OF VEGETATION AND LAND USES

Vegetative and bacterial activity in soils produces or consumes protons and is important in the development of soil acidity. Biologically mediated processes (Fig. 5.5; Tables 5.1 and 5.2) involving H^+ transfer are cation uptake, ammonium uptake, mineralization, N fixation, organic sulphur oxidation, and mineralization of organic phosphorus. Chemical processes also occur; they include dissociation of water and organic acids, mineral S oxidation or reduction, nitrification of NH_4^+ and NO_x, and complexation. Some processes have seasonal or temperature-dependent changes— the uptake of nitrate by growing plants consumes H^+ while death and decay liberate an equivalent H^+ flux. Prolonged drought leads to oxidation in organic horizons, with release of the acid, oxidized, sulphate when rain follows, while waterlogging will reduce S to sulphide, leading to a rise in soil pH. Freeze–thaw cycles also show an increase in nitrification and the release of protons. Land-use practices are also significant—deep ploughing breaks up iron-pan deposits in podzols, and helps to oxidize mineral sulphide or resistant organic materials; forest felling is followed by a surge of nitrate release from the rhizosphere. Even encouragement of permanent woodland or pasture increases the organic component of soils and their progressive acidification (Johnston et al., 1986).

Consideration of the relative magnitude of these processes led Rosenqvist to challenge the view that acidification was simply dependent on acid rain. Numerous studies, including budgets of ion transfers, now provide evidence of the important effects of tree uptake and harvesting in managed forests. Reasonable arguments can be presented to show that harvesting reduces the soil reserve of cations more than leaching due to acid soil conditions. Moreover, the presence of acid soils is not necessarily an indication of significant acid deposition, as shown by highly acid soils in unpolluted, pristine, forests in the northwestern USA, Australia or the Caribbean where nitrification, base uptake of cations or carbonic acid generation is demonstrated (Johnson et al., 1991). It is worth noting that CO_2 pressures in soil may be as much as 100 times greater within the soil than in the atmosphere and have the major weathering role.

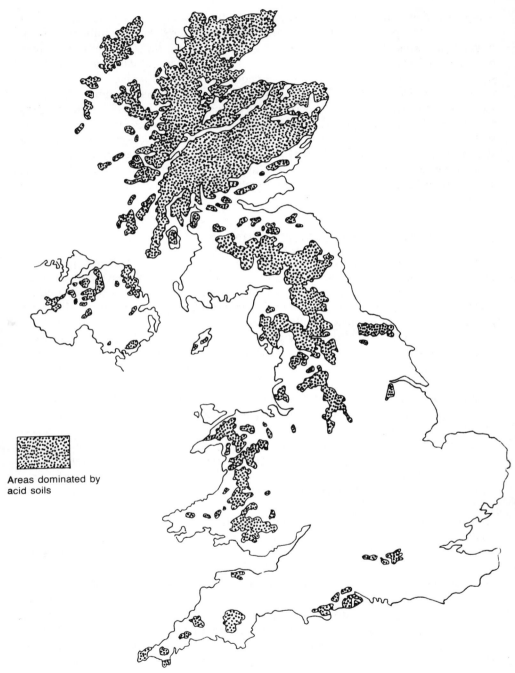

Fig. 5.4—Acidification sensitivity for the UK based on soil and bedrock characteristics
(after Warren *et al.*, 1986).

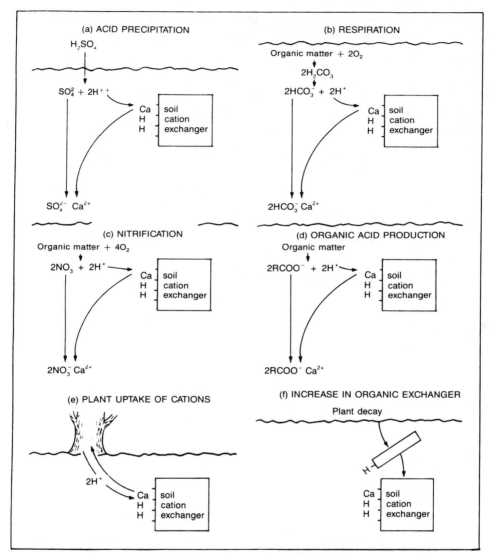

Fig. 5.5—Principal mechanisms of within-soil acidification (after Skeffington, 1987) involving soil exchanger mechanisms.

In some areas, however, acid rain appears to play a significant role. In cool temperate areas of high rainfall, such as the westerly regions of northern Europe and North America, soils are characterized by low bacterial activity, slow litter breakdown and the progressive accumulation of organic matter in soils (Jenkinson, 1970; Sollins *et al.*, 1980; Fig. 5.6). In these conditions, soil acidity is characterized by the presence of exhangeable Al^{3+} on clays and the association of H^+ and Al^{3+} from amorphous

Table 5.1—Balance of proton generation and consumption in terrestrial and aquatic systems (after van Breemen *et al.*, 1984)

Proton production	Proton consumption
Atmospheric deposition	Drainage from catchment
Cation assimilation	Cation mineralization
Anion mineralization	Anion assimilation
Dissociation of acids	Protonation of anions
Oxidation	Reduction
Reverse weathering of cations	Weathering of metal oxides
Weathering of anionic components	Reverse weathering of anions (anion retention)
Nitrification	Denitrification

Table 5.2—Estimates of acid (proton) production from various processes (after Mellanby, 1988; Skeffington, 1981; Farmer, pers. comm.)

Process	H^+ production ($keq\ ha^{-1}\ yr^{-1}$)	Comment
Rain, pH 5.0	0.11	At $1066\ mm\ yr^{-1}$
Acid rain, pH 4.3	0.86	Tillingbourne, UK[a]
Dry deposit	0.99	Tillingbourne[b]
Acid rain	2–7	Central Europe
	0.1	Northern Scandinavia
S oxidation	2.5	Oxidation of 2.7 cm peat
	0.1–10	In marine sediments
Vegetation uptake	0.04–5.0	Forest growth
	12–20	Arable farming (Scotland)
Respiration	6.1	Soil microbes
Carbon dioxide	10–20	On chalk substrate
	~0.5	On silicate rock
Nitrification	8.1	Soil microbes
	6–10	After clear felling
	High	After drought

[a]About $0.75\ keq\ ha^{-1}$ of strong acid, the remainder weak acids.
[b]$20\ \mu g\ m^{-3}\ SO_2$, deposition velocity $5\ mm\ s^{-1}$.
The figures are estimates from measured deposition and from field measurements of biological processes from the literature.

iron and aluminium oxides and organic matter (Johnson, 1987). The former is thought to be associated with acid rain and cation uptake, whereas the latter is affected by organic accumulation and weathering, affecting the weak acid component of soil acidity (Skeffington and Brown, 1986). At conifer-forested UK sites, runoff water is clearly more acid, and higher in aluminium, than in adjacent similar moorland areas (Adams *et al.*, 1990; Bird *et al.*, 1990; Hornung *et al.*, 1990b; Harriman and Morrison,

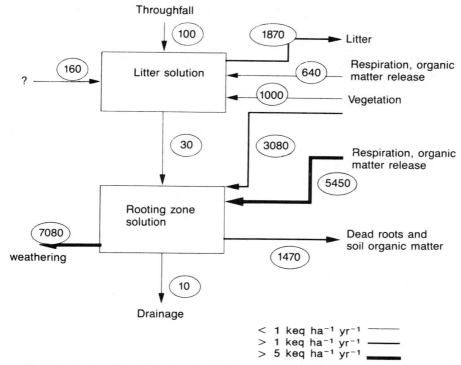

Fig. 5.6—Hydrogen ion (H$^+$) and cycling below mature Douglas fir, showing a predominant acid input from respiration and organic acid production (after Sollins *et al.*, 1980).

1982; Stoner and Gee, 1985). At all these sites, base-poor soils are depleted in cations and the soil exchanger is charged with H$^+$ and Al^{3+}, released by anion inputs in rain with a resulting surge in acidity and aluminium following rain storms. Less critical conditions are found at sites with mixed or deciduous trees. Pasture improvement techniques (ploughing, lime and fertilizer applications) also result in a runoff which is less acid and with higher calcium and magnesium and lower aluminium. Evidently, the quality of runoff could be improved by such land management practices.

In other areas, such as highly weathered soils of the southeastern USA, acidity is related to adsorption of anions and a high lime requirement—sulphate adsorption reduces the acidification due to base cation leaching but soil acidity due to anions is increased. In red alder forests in North America, microbial N fixation on the alder roots leads to N accumulation in the soil, with progressive soil acidification (doubling the N pool in 35 years), Al^{3+} mobilization, reduced base saturation and organic matter accumulation (Johnson *et al.*, 1991). The accumulation of N below alder over a 55-year period in Washington, USA, is calculated to match an annual input of 3.2 keq H$^+$ ha^{-1} in acid rain. In the Caribbean forests, carbonic acid is the major leaching agent and acid rain has little effect. In Australia, soils of native forests are often very acid due to the high uptake of minerals by the trees. Soil acidification is also reported in dunes planted with radiata pines which do not fix N, in comparison with adjacent eucalyptus stands—over 100 years, nitrogen in soil below the pines was much lower than that below the eucalyptus but pH was 1–2 pH units lower, associated with a much depleted base cation status of the soil. In studies elsewhere, for instance in eastern Canada, such consistent or substantial changes in conifer stands have not been found (e.g. Linzon and Temple, 1980).

In other unmanaged systems such as moorland, peat formation occurs and eventually thickens to a point where there is little effective exchange between upper and lower mineral horizons. Moorland ecosystems also accumulate nitrogen in litter, even though bryophytes eventually lose their capacity for N uptake from rain (Johnson *et al.*, 1991), and surface soils and runoff from peats become much more acid than the incoming deposition. There is still a question (Bache, 1993) whether the acidity of peats and their runoff is significantly affected by the acidity of rain, although some experimental acid leaching of peats suggests that organic soil acidity may rise by about 0.5 units. Paradoxically, the acidity in moorland runoff is enhanced if the rain contains significant sea salts. In these conditions the rain may be scarcely acidic since sea salt cations are present, but these are able to displace the H^+ and Al^{3+} accumulated and held by the soil exchanger.

The rapid rate at which some soils change suggests that the concept that soils reach some sort of steady-state condition is invalid. Tree uptake and recycling processes are dominant and uptake from deeper mineral soils and transfer through litter fall and foliar leaching to surface horizons may lead to major changes in the volume and chemical distributions between horizons of the soil profile, although this is less evident for the total soil profile (Fig. 5.7).

5.6 SULPHUR STATUS AND RETENTION

Some soils are able to absorb anions such as sulphate, usually on the positive surfaces of iron or manganese oxides (Johnson *et al.*, 1981; Singh, 1984). Retention of sulphate in soils is considered an important factor in the degree to which acidity or sulphate in rain is transferred to surface waters (Reuss and Johnson, 1986); absorption/desorption studies in a variety of soils suggest that many show a degree of irreversible absorption (Irving, 1991). On common soil minerals (e.g. haematite, goethite) it is so strongly bound that mobility within catchments will be slight (Turner and Kramer, 1992). In contrast, vegetation has only a limited capacity to immobilize sulphate, and in the bacterial biomass it is quite labile. On the other hand, nitrate is seldom retained, being highly soluble and seasonally in demand for biological growth.

Sulphate from rain input to soils causes an anion shift in soil drainage from a composition dominated by HCO_3^- and organic anions to one dominated by SO_4^{2-}. The increase in SO_4^{2-} is greater on an equivalent basis than the decrease in carbonate and organic anions, raising the ionic strength of the soil solution. This is the basis of the **mobile anion** concept (Wiklander, 1975/76; Johnson and Cole, 1980) by which soil cations (with aluminium) are lost to surface waters as a result of acid rain sulphate (Henriksen, 1980). In maritime locations, sulphate and chloride inputs are increased

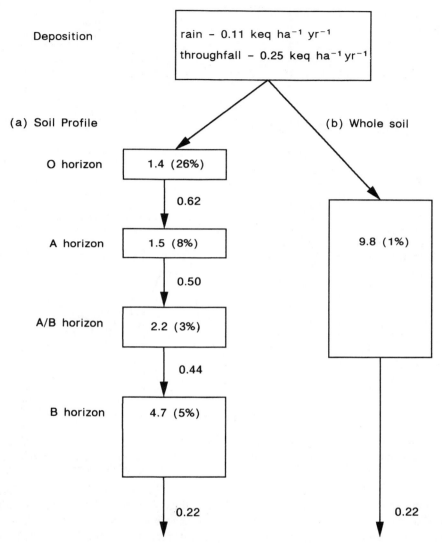

Fig. 5.7—Input/output budgets for magnesium for a mixed oak forest, calculated (a) as stepwise transfers for each horizon, or (b) treating the soil as a single entity (after Johnson *et al.*, 1991); values are keq ha^{-1} yr^{-1} and % values are for fraction of pool size.

by sea salts (Adams *et al.*, 1990; Ogden, 1982) and nitrate also, contributing significantly to total mobile anions even in upland and N-deficient soils (Cresser and Edwards, 1987).

Since sulphate in rain is often the major anion, it is often seen as a surrogate or partner for acidity in deposition, and as the signature of acid rain in surface waters. Early perceptions were that, at least on the basis of annual fluxes, sulphur deposited

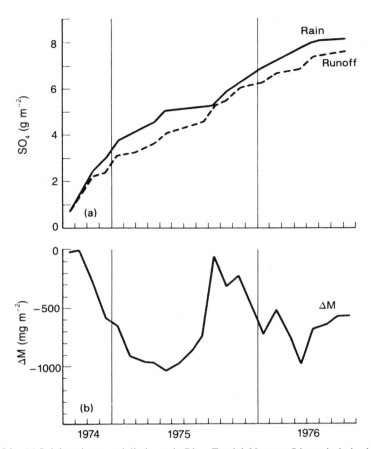

Fig. 5.8—(a) Sulphate input and discharge in River Tovdal, Norway; S input is derived from bulk rain, underestimating total deposition. (b) A seasonal change in sulphate retention (ΔM) is found from the difference between input and discharge.

in rain would be matched by an equivalent efflux in runoff (Fig. 5.8(a)). Although most of the sulphate in surface waters does indeed come from rain, the role of soils in retaining and releasing sulphate is important; recent quantification of S budgets in soils has improved our understanding. A relatively small amount of sulphur is taken up by the biota, to be recycled by decay, and an early attempt to match input and output (Dovland and Semb, 1978) showed a seasonal fluctuation in stored sulphur (Fig. 5.8(b)). In maritime areas receiving enhanced sulphate deposition from marine aerosols, a significant shortfall of efflux against influx provides evidence of S retention in catchment soils (Ogden, 1982).

It has also been demonstrated that sulphur deposited or drained to wetlands can be reduced to sulphide and retained (Brown, 1985; Bayley *et al.*, 1986), even to the extent of 70% of input (Fig. 5.9), reducing the acidity of drainage water. With a lowered water table during drought, the reduced S is reoxidized and released as sulphate.

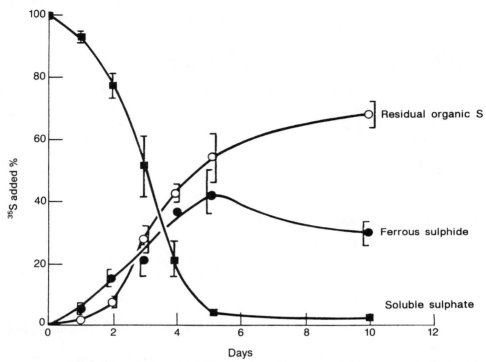

Fig. 5.9—Sulphur reduction in a valley mire; the course of S metabolism was tracked by radioactive ^{35}S (added as sulphate on day 0). Within five days sulphate has been reduced to sulphide, reducing acidity, and to organic complexes, both mechanisms effectively retaining S within the bog (after Brown, 1985).

Such sulphate releases follow autumn rains or snow melt, and the same phenomenon is shown in field experiments (Bayley *et al.*, 1986). In bogs where reducing conditions occur, the sulphate is reduced by microbial, as well as chemical, processes to FeS_2 if iron podzols are present (Hornung *et al.*, 1990b) or even to volatile sulphur (Brown, 1981). Similar processes occur in lake sediments during seasonal stratification of anoxic deeper waters, possibly reversed during lake overturn (see Chapter 6). This reduction of sulphate is accompanied by increased alkalinity.

In a major field experiment in Norway (RAIN, see Chapter 11) a fivefold increase in sulphate input at one site (Sogndal) led to only a doubling of the sulphate output, 80% being retained in the soil. At the other site (Risdalsheia), a 90% reduction of sulphate input has led to only a 40% reduction in output, some being retained by the soil S pool (Fig. 5.10). Easily soluble sulphate represents only 1–10% of the total S present (Brown, 1982), so some mineralization must be occurring, releasing long-term accumulated sulphur; total soil S at this site was calculated to be in excess of that accounted for by post-industrial deposition (Chester, 1986).

The ability of soils to retain sulphur has significant consequences for long-term acidification (Seip, 1980). An important distinction is to be made between areas where

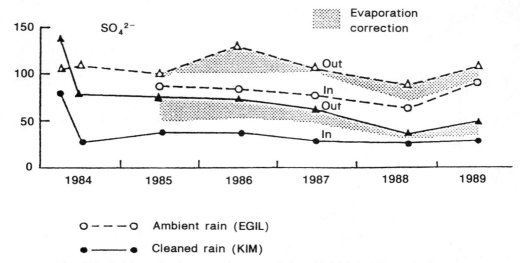

Fig. 5.10—Sulphur release over a six-year period at Risdalsheiea, Norway, where one catchment (——) receives 'cleaned' rain, while acid rain is applied to another (–––). The excess of sulphate discharge (shaded) is attributed to release of soil S (output is corrected for evapotranspiration) (after Skeffington and Brown, 1992).

soils are in a steady state with regard to sulphur, those exhibiting net accumulation, and those with non-reversible acidification (Rochelle *et al.*, 1987). In the first, reduced sulphur input generally results in lowered sulphate output in runoff (as at Sudbury, Canada: Gunn and Keller, 1990) although it is difficult to match the change in input to surface water changes with precision (Dillon *et al.*, 1986) since soil conditions are changed with lower acid input. In North America, a relationship between changes in input and runoff is found in areas north of the limit of Pleistocene glaciation (Galloway *et al.*, 1983; Rochelle *et al.*, 1987). In southern areas of the USA and in central Europe south of the limit of glaciation, sulphate concentrations in runoff are lower than expected from estimated input. Soils in these areas show substantial sulphate retention—50% in the Harz mountains, for example (Hauhs and Wright, 1988). In areas where deposition has been excessive, soils may have undergone major chemical or physical changes that prevent recovery by natural processes, except on geological timescales. In these soils, organic matter has been lost and aluminium leaching has led to the dissolution of clay minerals and the accumulation of heavy metals (Hauhs and Wright, 1988).

5.7   SOIL ACIDIFICATION: TIMESCALES OF CHANGE

The crucial questions are whether the greater acid deposition of the last two centuries has increased soil acidification and whether control of S and N emissions will reduce acid rain and encourage a reversal of soil acidification.

The preceding text will have shown that far from being stable and unreactive, most soils are in a state of change. Present-day soils in the northern temperate zone are

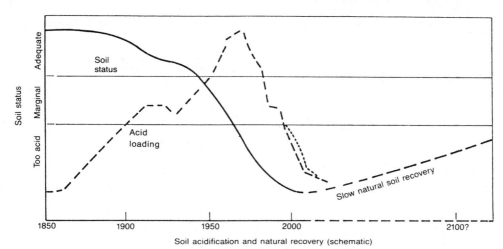

Fig. 5.11—Soil acidification over geological time—a schematic representation. Prior to industrial development, natural rain (pH ∼ 5) and organic soil development acidity depleted soil base reserves. After 1900 the greater acidity of rain accelerated the process. Recovery is dependent on bedrock weathering to replace lost cations. A slight improvement (·····) is expected with reduced S emissions.

the result of changes since the last glaciation, perhaps 20 000 years ago. The timescale of soil formation from the geological parent materials must differ according to their nature from several thousand to a few hundred thousand years, during which time natural processes will have lowered the pH and depleted the base reserves. Against this scale, the enhanced acid deposition over the past 150–200 years represents a slightly steeper decline in a long-continued process; this change in slope has significant implications for the predicted recovery of acidified ecosystems if deposition is reduced (Fig. 5.11; see Chapter 11). The pool of acidity in soil greatly exceeds that entering in rain; it represents the accumulation of decades or even millennia of change—from N fixation, sulphur deposition and organic matter, as well as H^+ deposition.

In the UK, many natural soils are considered to be sensitive to acidification because of the low base status of their parent bedrock. In Scotland, where the glacial ice sheet receded only about 10 000 years ago, estimates of weathering suggest that over that time all the calcium from 0.5 m depth of soil could be lost in an area in Scotland receiving 1100 mm of acid rain a year (Edwards *et al.*, 1985). In most soils mineral weathering would be adequate to maintain their base status, but weathering rates are notoriously difficult to estimate, and in the peaty conditions prevailing in much of Scotland, weathered materials may not be accessible. In contrast, in fresh Alaskan moraines, pH fell from an initial pH 8 to pH 5 within a few decades, first by the loss of carbonates, then by accumulation of organic matter and loss of base saturation (Johnson *et al.*, 1991).

The studies reported above on long-term accumulation of acidity in soils below forest stands provide a timescale for soil acidification, independent of an acid rain input. Remeasurement of soils within the same area over a period of time also provide

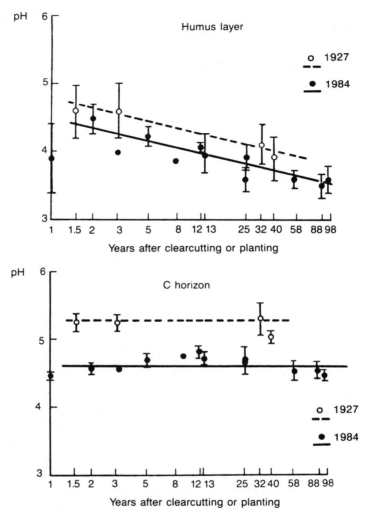

Fig. 5.12—Soil acidification of humus and C horizon in soils below a spruce stand over 40
years, as a log function of stand age. Acidity in the humus layer falls as the forest gets older,
but there is also a shift over the 40 years, especially in the C horizon (after Hallbacken and
Tamm, 1986).

insight into the rate of soil acidification (Johnson *et al.*, 1991). In the Adirondacks,
48 sites resampled after a 50-year gap showed that acid soils (pH < 4) did not acidify
further, although less acid soils did so to a small degree. At Walker Branch, Tennessee,
changes between 1971 and 1982 were found below deciduous trees, with depleted
base cations and CEC. The most notable evidence comes from southern Sweden,
where Tamm and Hallbacken were able to resample sites in 1984 first reported by
Tamm's father in 1927 (Fig. 5.12), using the same methods. Two trends were

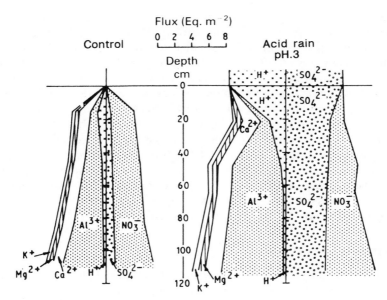

Fig. 5.13—Relative fluxes of major ions through two undisturbed soil columns watered for five years with simulated 'rain' at pH 5.6 and at pH 3. The width of the columns is proportional to flux at various depths. The acid lysimeter neutralizes the acid input, but the large flux of anions mobilizes Al^{3+} (after Skeffington and Brown, 1986).

identified—a pH decline in organic horizons with tree growth, and another in both A and C horizons independent of the age of the stand; surprisingly the decline is greater in the mineral horizon where cation reserves might be higher. The authors concluded that the acidification was a combination of acid rain and biological acidification (Hallbacken and Tamm, 1986). In another Swedish study, Falkengren-Grerup and Erikson (1990) reported changes in soils in a mixed forest and in pasture and heathland between 1949 and 1970; some acidification was evident, although methods differed. The loss of soil cations can be reasonably attributed to the uptake by trees, although a similar cation decline was seen in heathland sites.

In the absence of better historical data over sufficient time, experimental studies may hold more promise, although the long timescale of response expected at realistic rates of acid rain deposition is a practical problem with research programmes designed usually on a much shorter scale. In one study, where large undisturbed soil columns were subjected over a five-year period to a 1500 mm yr^{-1} 'rain' at pH 3, some acidification was evident from increased leaching of base cations, reduced base reserve and increased mobilization of aluminium, decreasing the amorphous, but not the exchangeable, aluminium in the soil (Brown, 1985; Skeffington and Brown, 1986; Fig. 5.13). Although the soils were subjected to the equivalent of 50–75 years of acid rain at the pH prevailing at the source site (pH 4.2), sulphate release was only 30% of the input and sulphur was retained.

In a large-scale field experiment (RAIN) in Norway, where two sites with contrasting deposition regimes were compared, the soils were initially only weakly acid. At Sogndal (soil pH 4.2–5.8) increased sulphate input was reflected in higher output in runoff, but with some S retention in soil. At Risdalsheia (soil pH .3.7–4.7) where soil base saturation was lower, after five years exclusion of sulphate it continues to be released to runoff; about 25% of sulphate in runoff prior to exclusion must have been derived by mineralization of sulphur stores in the soil (Skeffington, 1992). The decreasing sulphate is now being replaced by increased organic anions, as predicted by Krug and Frink (1983).

In another field 'experiment' resulting from the massive reduction (by 350%) of smelter emissions at Sudbury in Canada, lakes appear to have recovered quite quickly (Gunn and Keller, 1990), indicating a rapid reversal of soil acidification in this area where weathering rates are relatively high (Skeffington and Brown, 1992). In other less mineralized areas nearby, soil recovery may be slow because of lower weathering rates and the release of sulphur reserves in the soil.

5.8   UNRESOLVED ISSUES

Methods of soil analysis are not well standardized, especially for non-agricultural soils, so results from different projects cannot often be compared.

Soil heterogeneity on a small geographical scale makes sampling difficult and the variance of the data may make comparisons 'insignificant'; for practical reasons this variance can seldom be overcome by more samples. Although sizeable regions may be characterized as 'sensitive' on the basis of geology and soil, some small areas may not be protected by application of a critical load, while others may be unnecessarily protected.

Many natural processes, some seasonal or due to changes in land use, confound the possible effects of deposition changes (looking for a small signal against a large 'noise').

Sulphur retention and release from soils and wetlands are demonstrated, but are not well understood on a catchment scale, or over longer timescales.

The degree to which soil acidification might be reversed by reducing the acidity of rain is not established.

Cation-exchange processes and anion mobility are clearly critical processes, but there are still inconsistent observations and interpretations.

Experiments to exclude or enhance deposition have not yet resolved all uncertainties of interpretation; it is dangerous to extrapolate over space and time.

Overall, the degree to which soil acidification is attributable to natural phenomena, or to acid rain, is not clear. Timescales for soil acidification are equally uncertain.

5.9   SUMMARY AND CONCLUSIONS

The expectation that 'acid soil' generates 'acid water' is tenable, but that 'acid rain' generates 'acid water' is evidently not so in some cases; soils play a crucial role.

Upper soil horizons are more acid; in high rainfall areas, or in high flow conditions, most runoff has contact only with these soils.

Similarly, where podzol formation or peat growth restricts permeation of water through to the lower soils, contact is minimal.

In upper soils, organic residues and biological activity contribute to organic acids in runoff, while lower mineral horizons may provide weathered cations to neutralize acidity.

The stratified nature of soils and the variable pathways of drainage mean that hydrology is a critical factor determining runoff chemistry.

Cation exchange and mineral weathering play a significant part in neutralizing soil and runoff acidity. The increase in anions (and ionic strength) from rain, especially from maritime aerosols, provides transport for soluble aluminium.

Vegetation plays an important role in moving cations from the soil into biomass, where they are sequestered until decay recycles them. Cropping removes them from the catchment.

In some areas, where vegetative growth or microbial activity is large, there is rapid soil acidification in the absence of acid rain.

In more temperate areas, there is slower accumulation of organic materials and progressive but slower soil acidification even on alkaline soils. Acid rain has also contributed to acidification.

Soils have been subject to progressive changes since soils were laid down in glaciated areas on a geological scale. If weathering fails to match uptake of cations by vegetation and their leaching to runoff, the soil will become acidified.

Sulphur retention in soils varies with soil properties and previous glaciation, as well as accrued S deposition, and will influence the response to changes in deposition.

6

Surface water quality

6.1 INTRODUCTION

What is an acid lake? Where do they occur? How many are there? These simple questions are not easy to answer. Natural waters differ in their characteristic physical, chemical and biological properties and so in the degree and sensitivity to acidification. Although each is unique, it is possible to make some valid generalizations, specifically on the processes that lead to acidification and to some extent to its reversal.

The great majority of the world's waters are circumneutral or slightly alkaline in character (Alabaster and Lloyd, 1982; Rodhe, 1949), even though they drain catchments of varying bedrock, overlain with diverse soil types, and subject to different cultural and management systems (Fig. 6.1). In fact, more than 97% of fresh waters discharging to the oceans are buffered by the carbonate/bicarbonate system, with pH > 5.6. This almost universal similarity stems from the weathering of minerals to potentially soluble forms which provide the solutes in runoff where their concentrations are governed by solute equilibria. These are influenced by temperature, atmospheric conditions, pH and ionic strength. Local or regional differences in rain quality play a relatively minor role, although rain quantity (determining the contact time with soils) and the input of marine salts in maritime regions are significant. Ionic strength (measured as conductivity, units $\mu S\ cm^{-1}$) of most waters is usually substantial, up to $10^3\ \mu S\ cm^{-1}$ in high hydrological-order rivers or large lakes with long retention times; the solutes present provide a significant buffer capacity that helps to stabilize the acid–base balance (Rodhe, 1949).

There are exceptions to this general picture, notably the high-conductivity alkaline 'soap lakes' found in some desert regions, those influenced by volcanic activity with high levels of sulphate or those on anomalous geological extrusions. There are also the low-conductivity acid lakes, often streams of low hydrological order or lakes with short retention times, draining areas of unreactive geology of slow weathering rates. Flows are usually rapid and associated with high rainfall. The range of reported pH values world-wide extends from pH 1.7 and 1.8 in volcanic lakes to pH 12 in closed alkaline lakes. Most productive lakes are in the range pH 6 to pH 8 (Howells, 1983).

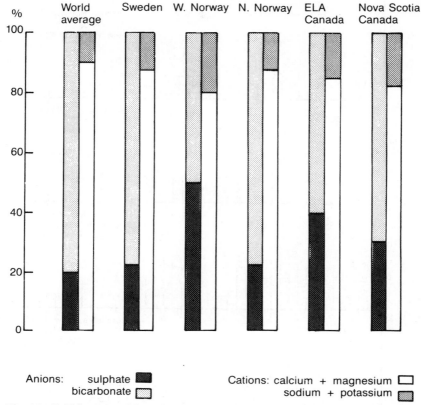

Fig. 6.1—Relative composition of major non-marine ions (as % of anions or cations) of some surface waters (after Henriksen, 1980); note the differing balance of sulphate and carbonate; non-marine chloride is considered zero.

On a global basis acid waters are the exception, but they are numerous in some northern temperate regions such as Scandinavia and parts of North America. They are also found in the Amazon, Japan, Australia and New Zealand. They include waters draining wetlands, especially with *Sphagnum* moss (pH 3.3 to 4), and other 'blackwater streams' of high humic content, such as in the Amazon basin (pH 2.8–5.8). Some glacial lakes, or those formed on recent morainic materials, receiving rain and meltwater with little weathered material, may have pH values down to 3.3. Lakes and streams of coastal sand plains (Queensland, Australia and New Jersey, USA) which have exhausted their stock of weatherable cations, have pH values of 4.2 to 4.5.

The lakes and streams generally regarded as 'acidified' (i.e. more acid than they would be by natural causes) are the ultra-oligotrophic waters (about 50 μS cm^{-1} or less) draining areas underlain by unreactive geology such as granites, gneisses, quartzites, or sandy soils and schists (Fig. 6.2). In these areas the supply of weathered cations is not sufficient to counter the acidifying input of rain, soil acid-generation

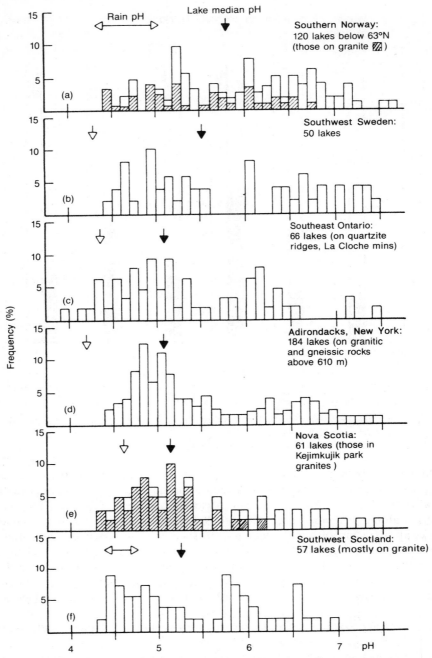

Fig. 6.2—Distribution of pH in groups of acid lakes, reflecting differences in geology and annual rain pH.

processes and uptake of cations in vegetative growth. Such lakes and streams are found on the granitic areas of southern Norway, southwest Scotland, the Adirondacks (USA) and the Canadian Shield, the thin morainic podzols of southern Sweden, and the schists, mudstones and sandstones of Wales.

6.2   CHEMICAL CHARACTERISTICS OF ACID WATERS

Acidity is conventionally considered in terms of 'pH', the negative logarithm of the hydrogen ion concentration (i.e. $[H^+]$), each unit of pH representing a tenfold difference in hydrogen ion concentration (Fig. 1.2). As hydrogen ion concentration increases, pH falls: at pH 6.0, $[H^+]$ is 1 μeq l^{-1}, and at pH 5.0, $[H^+]$ is 10 μeq l^{-1}. Reported values of acidity in surface waters range from pH 1.3, $[H^+] > 10\,000$ μeq l^{-1} in some volcanic lakes, while in some concentrated, non-draining, 'soap' lakes with pH > 12, $[H^+]$ is absent and hydroxyl ions are present with $[OH^-] > 1000$ μeq l^{-1}. Strictly, any lake below pH 7.0 could be termed an acid lake, since protons (H^+) will exceed hydroxyl ions (OH^-). In practice, acid waters are measurably more acid, but various arbitrary thresholds are found in the literature; some argue that acid waters are those of pH < 6 or < 5.0, or < 5.6 where there is equilibrium with atmospheric CO_2 and distilled water (at atmospheric pressure and 20°C), or < 5.7 where most carbonate alkalinity is lost. These limits are not necessarily justified by the chemistry or biology of the water body. Acidic waters were defined by the NAPAP programme as those with a zero acid neutralizing capacity (ANC, see below) (L.A. Baker et al., 1990), but elsewhere (Henriksen et al., 1992) as those with < 50 μeq l^{-1}. In some cases, a biological criterion is used, with pH < 6.5 (i.e. $[H^+] < 1$ μeq l^{-1}) since some classes of organisms may be absent (Raddum and Fjellheim, 1984) or pH < 5.7 where the benthic community is impoverished (Sutcliffe, 1983), or where 'circumneutral species' (diatoms) are absent. Thus, the term 'acid water' is somewhat subjective and specific to the case being made.

In practical and realistic terms, it is important to note that natural waters are seldom in equilibrium or uniform within a water body, particularly so for unbuffered waters of low ionic strength. For lakes, a mid-lake sample taken in autumn after break-up of any summer stratification may provide a representative value, while for streams, with substantial longitudinal as well as seasonal variation, a single value cannot be characteristic. Field or laboratory measurements are typically within ± 0.2 pH units standard error (Gardner, 1985); values obtained are dependent on the method of measurement, temperature, sample handling and whether it is 'open', i.e. equilibrated with atmosphere, or measured in a closed system. Because of the log transformation of H^+ concentrations, pH changes around neutrality may seem large (say 1 pH unit) but represent only small differences in $[H^+]$ and may be biologically trivial, as well as inaccurate. In contrast, a similar unit change in pH, say from pH 5 to pH 4 representing an increase of 90 μeq l^{-1}, is obviously of greater ecological significance.

A particular problem with regard to the chemical characterization of these waters is that they are often of low hydrological order (near to source) in areas of high rainfall with 'flashy' flows and a rapid lake turnover—features associated with highly variable chemistry. Their chemical nature cannot be assessed with confidence from 'spot', occasional or random samples. Natural variations in water pH are often much larger than those associated with deposition (Turk, 1985). This problem is now usually recognized, but it follows that much of the early data reported can be accepted only with reservations (Kramer and Tessier, 1982), especially if the protocol of sampling and analysis is not explained. A further analytical problem arises from the very dilute nature of many acid waters, since techniques of suitably sensitive analytical methods have only recently become widely available. This is particularly true for pH; earlier colorimetric methods often used in the field employed the addition of unrecorded quantities of colour reagents of high ionic strength which would influence the acidity of the low conductivity waters being tested. Even modern electrometric techniques are subject to uncertainty because electrodes are not produced to a uniform or standard sensitivity and are not always calibrated to a realistic chemical standard. One UK data set over a ten-year period showed a sharp change in pH when an automated addition of buffer and indicator replaced the previous manual method; the pH difference was as much as 2 pH units (Ellis and Hunt, 1986). Even when the same samples are measured in the field and then in the laboratory by the same techniques, the values usually differ, partly as a consequence of equilibration and other changes which occur with time in unbuffered water, partly due to change in the conditions of measurement which may result in a variable equipment response (Kramer and Tessier, 1982). In a few instances, colorimetric and potentiometric techniques have been made in parallel, in an attempt to calibrate historical data; in a series of Adirondack lake samples, colorimetric methods overestimated pH, particularly at low pH, so that historical records would need correction for assessment of a temporal trend (Pfeiffer and Festa, 1980; Fig. 6.3).

Measurement of other chemical constituents may also be difficult, especially where concentrations are low. This is particularly true for phosphate and trace metals which have significant biological significance. It is also now recognized that many former methods for measuring sulphate were imprecise in the presence of dissolved organic materials (Kerekes *et al.*, 1984).

More credibility rests with chemical data sets derived from consistent and regular sampling at a specified site over at least a year where pH records are also accompanied by analysis of major ions, and alkalinity, preferably 'weighted' to take account of flow variations. For assessment of temporal trends, gaps in the record limit the application of most statistical techniques; the time period required for statistical significance will depend on the variance of the data, and whether the change is regular or random.

Acidity and alkalinity

Neutral waters, following the definition of Sorenson, are those where $[H^+] = [OH^-]$, giving a pH 7.0 at 25°C. The acidity of runoff and surface waters is expressed as an excess of $[H^+]$ reflecting a deficit of basic cations:

$$\{SA > \{SB$$

where {SA and {SB are the sums of strong acid and base cations. However, pure water in equilibrium with carbon dioxide in atmosphere has a pH 5.6 (see Chapter 1) and this pH might serve as a standard against which the acidity of rain or surface

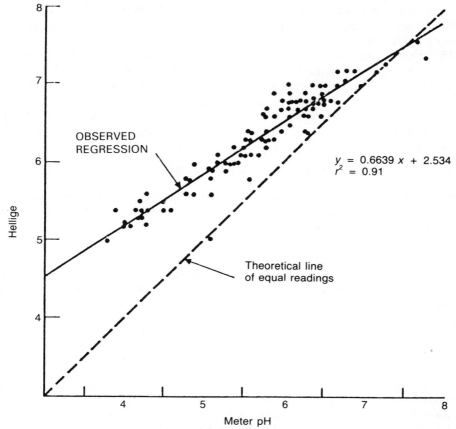

Fig. 6.3—Colorimetric (Hellige) versus electrometric measurements of pH; comparative data
for Adirondack lakes (after Pfeiffer and Festa, 1980).

waters might be measured (but see Chapter 1). However, surface waters have a higher
content of solutes and may be affected by photosynthesis or respiration, or the
production of CO_2 within soils (up to 4% v/v; Skiba and Cresser, 1991) and not in
equilibrium with atmosphere. Other somewhat arbitrary 'definitions' of an acid water
are:

$$[SO_4^{2-}] > [HCO_3^-], \quad \text{or} \quad [SO_4^{2-}] > [Ca^{2+} + Mg^{2+}]$$

but these have no strict scientific basis and ignore the influence of many other
constituents of natural waters.

   In addition to 'strong' (mineral) acid anions such as sulphate, nitrate and chloride,
some 'weak' acidity may also be contributed by organic acids such as humic or fulvic
acids generated by bog vegetation, or phenolic or acetic acids from vegetation decay.
These weak acids alone are able to depress the pH of low ionic strength waters to
around pH 5.0 owing to their strong carboxylic functional groups (Kramer et al.,

1989). At pH < 4.0, the strong mineral acids are thought to predominate (Henriksen, 1980). Organic materials also influence many chemical reactions and the potential toxicity of trace metals.

The acidity of runoff and surface waters is typically less than that of incident rain, due to the solution of neutralizing cations and/or reduction of sulphate in the soil (see Chapter 5). However, in areas where peat bogs or organic soils occur, the drainage waters may be more acid than rain, even by an order of magnitude in $[H^+]$ or > 1 pH unit; at Loch Fleet in southwest Scotland, rain pH is 4.8, lake pH 4.0–4.4 (Howells and Dalziel, 1992), and in the Amazon basin, rain pH is 5.0, stream pH 2.8.

Alkalinity is the obverse of acidity; it is expressed chemically as:

$$Alk = [HCO_3^-] + 2[CO_3^-] + [A^-] + [OH^-] - [H^+]$$

where A^- represents the contribution of weak acids. At pH 5.6 to 9.0, carbonate, hydroxyl and hydrogen are all small in comparison with bicarbonate (HCO_3^-), and the equation can be simplified to:

$$Alk = [HCO_3^-] + [A^-]$$

Alkalinity, like acidity, is an *intensity* (concentration) factor, and should be distinguished from **acid neutralizing capacity** (ANC) which is defined as the difference between the strong base anions and strong acid anions:

$$ANC = [Ca^{2+} + Mg^{2+} + Na^+ + NH_4^+ + TM] - [SO_4^{2-} + NO_3^- + Cl^- + F^-]$$
$$= \{SB - \{SA$$

where TM represents cationic trace metals. A similar base neutralizing capacity (BNC) could be defined, but is seldom used. Alkalinity can thus be estimated from measured major ions, but commonly from carbonates ('carbonate alkalinity') measured by direct titration with a strong acid ('Gran titration') (Mackereth *et al.*, 1989; Stumm and Morgan, 1981). Non-carbonate ANC can be defined as:

$$ANC_{nc} = Gran\ ANC - Alk$$

Historical Gran ANC measurements were usually to a fixed end-point using a colorimetric response—e.g. to pH 4.5 with methyl orange indicator—which gives an overestimate of alkalinity to a maximum of about 30 μeq l^{-1}. More recent studies use the plot of alkalinity remaining against acid added to measure a 'negative alkalinity' (Henriksen, 1982c; Fig. 6.4). Because the direct titration method is time-consuming, an empirical correction factor for Norwegian lakes from a sample of 67 lakes was used by Henriksen to provide an 'equivalence alkalinity' (Alk-E):

$$Alk\text{-}E = -1.9 + (1.00 \times SA)$$

where SA is the concentration (in equivalents) of strong acids, a correction considered important in waters of pH 4.0 to 4.5. However, it applies to the Norwegian waters sampled, and should be determined specifically for other waters if used. Many data sets for the UK do not fit the Henriksen relationship for pH and alkalinity (AWRG, 1987), possibly because simplifying assumptions (e.g. dominant bicarbonate buffering)

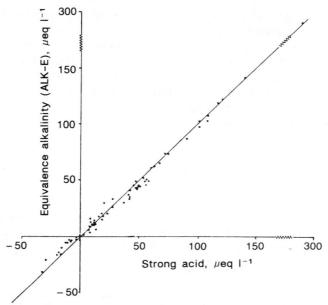

Fig. 6.4—Chemical relationship of equivalence alkalinity and strong acids (after Henriksen, 1982c).

were not applicable to UK waters. A further generalization reached by Henriksen on data for 13 unacidified lakes world-wide was that:

$$Alk = 0.93([Ca^{2+}] + [Mg^{2+}]) - 14$$

This was assumed to represent a universal pre-acidification alkalinity (Alk_0) against which the degree of present acidification could be judged. This concept is explored further below.

Surface water anions

A complete list of anions present in runoff and surface waters is probably not possible, since organic anions are seldom identified precisely, and some anionic forms of elements such as aluminium will vary with pH and other conditions. Generalizing, acid components are:

$$2[SO_4^{2-}] + [NO_3^-] + Y[Org] + [Cl^-] + [HCO_3^-]$$

Since chloride is considered to be present only as neutral sea salt, nitrate is assumed to be negligible and aluminium is absent in non-acidified waters, the important acidifying anions reduce to sulphate and bicarbonate. It is also common to calculate the non-marine component of sulphate (often shown as $SO_4^{2-}*$) on the basis of chloride present. These simplifying assumptions are not always realistic. Marine chloride contributes about 14% of world average river concentrations (Li, 1990) and may be the major anion present in maritime areas, and along with sodium, plays a role in soil exchange and biological processes; both may also be contributed by geological sources. Similarly sulphate from marine sources and soil reserves, as well as that from rain with current sulphur emissions, is reflected in total runoff concentrations. Furthermore, at least some is derived from decay and other biological processes (see Chapter 5). Nor can nitrate be ignored, especially where man-made sources enhance natural processes of nitrification in soils; outside the growing season in temperate latitudes, it may build up in soils, and if present in excess of that needed for growth, will 'spill' to increase nitrate concentrations in runoff.

Surface water cations
A similar list of cations present would be:

$$[H^+] + [NH_4{}^+] + 2[Mg^{2+}] + 2[Ca^{2+}] + [Na^+] + [K^+] + x[Al^{3+}]$$

Again ammonia is usually assumed to be negligible, potassium very small and the component of aluminium as Al^{3+} dependent on water quality. Sodium, like chloride, is often ignored and so a simplification could be:

$$\Sigma Cats = [H^+] + 2[Ca^{2+} + Mg^{2+}]$$

Again these assumptions may not be acceptable in conditions where sea salt or geological influences are strong. In oligotrophic waters, calcium and magnesium concentrations are typically low—a few milligrams per litre—and often below the mineral requirement of many aquatic animals, and marine magnesium may be a significant source; their supply is from rain (including sea salts) and the weathering of minerals in the soil. Ammonia is a significant component of rain in areas of intensive agriculture and adds to the burden of acid deposition since it is freely oxidized to nitrate. While almost completely absent in unpolluted surface waters, especially oligotrophic ones, it does sometimes occur during lake stratification or prolonged ice cover where anoxic conditions prevail—a condition that has been associated with 'winter-kill' of fish populations in North America (Casselman and Harvey, 1976).

Major ions in typical oligotrophic acid waters may be considered as the balance between anions and cations:

$$[H^+] + [Na^+] + [K^+] + 2[Ca^{2+}] + 2[Mg^{2+}]$$
$$= [Cl^-] + 2[SO_4{}^{2-}] + [NO_3{}^-] + [HCO_3{}^-] + [OH^-]$$

ignoring both cationic and anionic aluminium species, organic species and other trace components. Accepting the simplifications above:

$$[H^+] + [Ca^{2+} + Mg^{2+}] = 2[SO_4{}^{2-}]$$

or

$$[SO_4{}^{2-}] - [Ca^{2+} + Mg^{2+}] = [H^+]$$

Southern Norwegian lakes sampled by Wright and Snekvik (1978) (see also Wright and Henriksen, 1978) and the southern Norwegian Tovdal river (Dovland and Semb,

Table 6.1—Characteristic lake chemistry in Europe, North America, and Australasia in low and high acid deposition areas. Composition (μeq l^{-1}) reflects catchment conditions as well as human influence

Region	*SO_4^{2-}	NO_3^-	*{Cb	DOC	%ANC $\leqslant 0$
Low deposition regions:					
World av.	210	—	1200	—	
C Sweden	40	—	224	—	
N Finland	37	—	186	5.2	0
C Norway	20	11	44	1.4	15
US Rockies	25	0.4	154	1.3	<1
ELA, Canada	76	—	177	—	0
N Atlantic, Canada	32	—	83	5.7	3
Tasmania	47	—	429	12	3
High deposition regions:					
S Finland	140	2.4	354	12	8
S Norway	79	0.4	123	1.8	3
SE Ontario	150	—	192	4.2	5
Nova Scotia	124	—	616	5.7	0
Adirondacks	118	0.9	238	4.2	17
New England	136	0.7	533	4.0	6
Some other waters:					
Mine (USA) drainage	3272	20	2720[a]	1	—
Volcanic lake, NZ	4805	—	187[a]	—	—

*SO_4^{2-} and *{Cb (sum of bases) are corrected for sea salts in coastal regions; %ANC is proportion of lakes with no ANC, implying a pH < 5.3.
*[a]{Cb in mg l^{-1} for these waters.

1978) seemed to conform with this, viz. that sulphate alone was matched to the cation balance. This rather reductionist approach was used to simplify the concept of lake acidification, especially its link to acidity and sulphate in rain. It provided little guidance to identifying the causal mechanisms of acidification or for understanding relevant biological processes. Indeed, nitrate and ammonia together, on an equivalent basis, may exceed the level of non-marine sulphate in rain even in Norway (Henriksen and Brakke, 1988). Recent acknowledgement of the growing level of N emissions, nitrate and ammonia in rain, and nitrate in surface waters means that nitrate must also be included. It also ignores the significant effect of organic materials and aluminium or other trace metals, and the interrelationships between ions that are often significant for aquatic life (see Chapter 8).

Some examples of major ion chemistry in acidic and non-acidic waters are given in Table 6.1; significant differences between the level of total solutes and the various ionic components reflect the geology and soils of the drainage basins, land use and vegetation, and distance from the coast, as well as the input of rain. Even within the class of 'acidified' waters this leads to a wide range of water conditions.

Table 6.2—Water quality of some acid waters mentioned in the text. Even within a group of acid waters, chemical components vary widely

Lake	pH	SO_4^{2-}	NO_3^-	Cl^-	{Cb	DOC	Al^t	Al^i
SW Sweden								
Gardsjon	4.65	160	7	258	124[a]	2.2	290	
S Norway								
Birkenes	4.54	149	7	132		3–11		135–340
SW Scotland								
Fleet	3.99	141	8	264	308	15	220	100
					62			
NE USA								
Woods L.	4.7	126	19	9	200	2	14	
Nova Scotia								
15 m Brook	4.97	62	—	141	252	12	152	
Mersey R.	5.2	69	—	127	206	6.9	164	36

Ions are given as $\mu eq\,l^{-1}$, uncorrected for sea salts; DOC is $mg\,l^{-1}$; Al^t is total, Al^i is inorganic aluminium, $\mu g\,l^{-1}$. Fleet, unlimited system.
[a]Ca + Mg only.

Minor ions and organic materials

Reference has been made already to the mobilization of soil aluminium which is transferred via runoff to surface waters. A number of other trace metals, specifically iron, manganese, cadmium, copper, lead and zinc, are typically found in soft acid waters, usually in the (II) oxidation state at pH < 5 (Stumm and Morgan, 1981; McDonald *et al.*, 1989). They are derived from soils and bedrock, especially in areas of mineral ores, and like aluminium, are mobilized by acid soil conditions. Most are toxic to fish at low concentrations, but possibly have additive or even synergistic effects. On the basis of LC_{50} measurements for fish, their relative ranking is:

$$Cu > Cd > Zn > Pb > Al > Fe > H^+ > Mn > Ni$$

Copper, zinc, aluminium and iron are reported in excess of the LC_{50} values in a number of soft acid waters, but the degree to which they might explain fish declines is not clear. Lakes near the Sudbury smelter (Canada), where trace metals are high, have become less acid since emissions fell after 1970, but are still not suitable for fish, perhaps because of the residual levels of toxic metals. Toxic metals in acid mine drainage or in runoff from mine wastes is well recognized and researched (Kelly, 1988). Deposition of heavy metals from the atmosphere is less significant, except where there are uncontrolled emissions nearby. In Finnish headwater lakes, trace metal concentrations and bioaccumulation in biota increased with greater acidity and lower ANC (Iivonen *et al.*, 1992) but accumulation was lower in lakes with high organic content.

Dissolved organic material is substantial in some drainage waters and its role in acid waters is acknowledged by considering it in ion balances. Organic acids are also important weathering agents in soils (Jacks, 1990). In the northeastern USA and southern Norway, dissolved organic carbon (DOC), an indicator of organic acids, is not closely correlated with acidity or ANC, and is considered unimportant; the levels of DOC are usually $< 2 \text{ mg l}^{-1}$. Elsewhere, as in Finland, organic acids, however, may contribute as much as 60% of stream water acidity at about pH 5.0 (DOC about 12 mg l^{-1}), and they may have an important effect in chelating potentially toxic agents such as aluminium. This is the case in Nova Scotia rivers with high total aluminium, reducing the toxic aluminium fraction to $< 7\%$ of the total (Lacroix and Kan, 1986; Lacroix, 1992).

6.3 RELATIONSHIP TO PRECIPITATION

It has been claimed that present-day composition of oligotrophic surface waters is largely determined by sulphur (and acid) deposition to the catchment; as a consequence acid waters are found in sensitive areas with 'unreactive' geology receiving acid rain of pH < 4.7 (Wright and Henriksen, 1978, 1979). As explained above, the acid conditions reflect the catchment's capacity to generate alkalinity sufficient to match acid inputs from whatever source. Regions identified as 'sensitive' to acidification are southern Scandinavia, western Britain, and northeastern America. However, not all lakes in these regions are acid, and conversely some are found in locations where rain is not so acid (i.e. pH > 4.7). In some instances an almost equivalent relationship between rain and lake sulphate concentrations was found (Wright and Henriksen, 1979) or at least a linear relationship with wet deposition influx (Fig. 6.5, L. A. Baker *et al.*, 1990). Since rain sulphate underestimates total S deposition, and since most soils retain sulphur to some degree, these relationships are uncertain and differ between regions, although they have been used to quantify the expected benefit in surface water acidity from emission control.

Nonetheless, as sulphate in rain increased between 1950 and the mid-1960s in Sweden, Oden and Ahl (1972) reported an increase in sulphate concentration in 15 rivers over the period 1920 to 1972, associated with increased acidity and decreased bicarbonate alkalinity; over this period the estimated runoff sulphate doubled, and pH declined by 1 to 1.5 pH units. In some other regions, even greater changes were reported. More recently, Jacks and Knutsson (1982) argued that at least some of these changes reflected short-term changes in climate, with less than normal precipitation during the 1970s. The lowered water table allowed soil sulphides to oxidize to sulphate and then to leach to groundwater. Alkalinity also declined during the dry years, to rise again during wet years, attributed to ground moisture preventing the diffusion of carbon dioxide from soils to atmosphere. Re-analysis of the Swedish data (Sanden *et al.*, 1987) has also shown no evidence that surface water quality changed significantly in the late 1960s, even though sulphur deposition doubled in the following decade, but has fallen since. Water chemistry in some lakes in southwest Sweden, however, suggests a fall in acidity and sulphate from about 1976 to 1978 (Forsberg *et al.*, 1985).

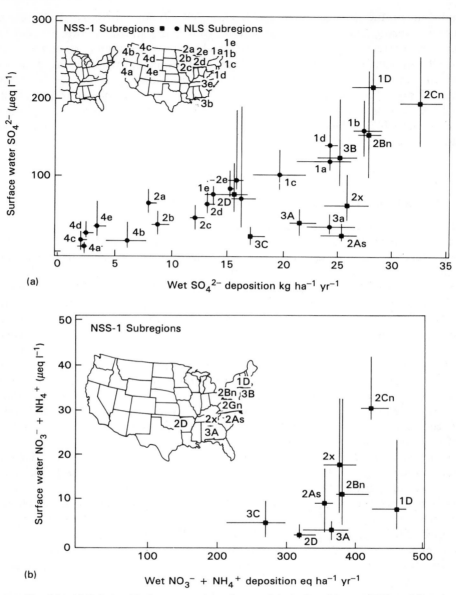

Fig. 6.5—(a) Relationship between estimated wet sulphate deposition and lake sulphate concentrations—note different regional relationships; (b) relationship for nitrate (after Irving, 1991).

Some specific sites

On a more site-oriented basis, relationships are even less clear. In two New Jersey streams studied over a 10-year period (1968–78) (Johnson, 1979), a parallel change in acidity and sulphate was found, but nine other streams in the same area did not follow the same trend, although the frequency of acid events did increase. Rain was acid in 1978 (pH about 4.0) but there were no historical data to link the stream quality changes to a change in rain chemistry. Streams in this region (New Jersey Pine Barrens) have significant organic content, although inorganic acidity dominates; about half have ANC $\leqslant 0$.

In Halifax, Nova Scotia, in a study of 12 lakes on granite or metamorphic rocks during the period 1955 to 1977, with doubling of local emissions, acidity and sulphate both increased by 16 μeq l^{-1}, although only that for hydrogen ions was significant (Watt *et al.*, 1979). The changes in acidity and sulphate, however, show little correlation ($r^2 = 0.11$) and a negative slope, suggesting that they are unrelated. In a study of rivers in the same area, Watt *et al.* (1983) attributed 73% of the acidity to increased sulphate, the mean change over 25 years being 9 μeq l^{-1} and 39 μeq l^{-1} for acidity and sulphate respectively.

Lakes near the giant smelter at Sudbury, Ontario, Canada, provide an opportunity to observe changes in chemistry following the massive decrease in local emissions and deposition since 1970 (Dillon *et al.*, 1987; Wright and Hauhs, 1991). In Clearwater Lake and Swan Lake, substantial changes are recorded, with sulphate and base cations falling and alkalinity increasing. In more distant Muskoka (southern Ontario) where sulphate deposition decreased by 30%, Plastic Lake sulphate concentrations scarcely changed overall, but acidification increased with falling alkalinity and base cations, attributed to a reduced supply of catchment cations and to stored sulphate. Harp Lake nearby has scarcely changed in its chemistry. It is inferred that acid deposition continues to acidify the soils, maintaining lake acidity.

Studies at Hubbard Brook, New Hampshire, provide a long-term and continuous data set of precipitation and stream chemistry since 1965. Sulphate in deposition has declined over more than 20 years by about 20% but has been ineffective in reversing soil acidification there, since stream pH is unchanged and base cation concentrations have fallen while sulphate has declined (Wright and Hauhs, 1991).

In the UK, there is little evidence (Warren *et al.*, 1986) that the upper reaches of monitored rivers are acid or have changed significantly over a decade. However, there is evidence from two acid upland tarns in southwest Scotland (see Chapter 7, Fig. 7.8) that acidity has been reduced (by < 0.2 pH units) in parallel with reduced (25%) sulphate deposition (Battarbee *et al.*, 1989; Wright and Hauhs, 1991). However, at two other Galloway sites nearby where continuous records are available over a decade, little change is seen in rain pH or sulphate, or in the sulphate of surface waters (Lees and Farley, 1993; Howells, 1991).

In Cumbria, stream acidity and sulphate have declined in streams of the Duddon catchment over a 10-year period (1970–80), but while sulphate concentrations have declined by 40–50 μeq l^{-1}, nitrate has increased by about 20 μeq l^{-1}; the associated acidity change is an increase of 13 μeq H$^+$ l^{-1}, while yields of calcium from catchment soils are unchanged (Tipping, 1989). A similar match in deposition and water chemistry has not been reported for other parts of Britain, although national emissions continue to decline (Warren *et al.*, 1988).

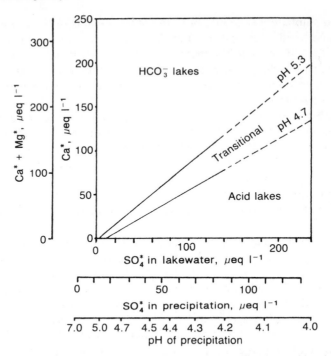

Fig. 6.6—Categorization of lakes in Norway: bicarbonate buffered, transitional and acid, with Henriksen's nomogram to estimate lake chemistry from rain (after Henriksen, 1979, 1980).

6.4   INDICES OF ACIDIFICATION

The degree to which waters are acidified is of interest, both to judge the extent and timescale of change, and to predict their recovery that might be expected from reduced deposition. An early classification of lakes (Fig. 6.6) as 'bicarbonate' lakes (pH > 5.6), 'acid' lakes (pH < 4.7) and 'transition' lakes of variable and unstable pH (Henriksen, 1979, 1980) was largely empirical. Their calcium and magnesium concentrations were plotted against the acidity and sulphate concentration of rain (Henriksen, 1980). In these oligotrophic Norwegian waters, it was perhaps not surprising that sulphate acidity in the water was matched by the sum of the base cations (as explained above), nor that the ratio of alkalinity to sulphate is reflected in the pH, even though the relationship with rain is more obscure.

Henriksen (1982b) further developed an earlier concept (Almer *et al.*, 1978) that the sum of calcium and magnesium represents the alkalinity of non-acidified waters, implying that a deviation from a 1:1 relationship is an index of the degree of acidification. The Henriksen nomograms provided a visual illustration, but later data show the importance of other acid and alkalinity components. A major shortcoming of the empirical approach was also that it cannot be used predictively without knowledge of how each catchment responds to changes in rain input, particularly with regard to its yield of calcium and magnesium. Further, some of the variables, viz. total organic content (TOC), SO_4^{2-} and Ca^{2+}, are clearly not independent of each other (Krug, 1988).

In its first formulation, Henriksen's 'titration' model assumed that $[Ca^{2+} + Mg^{2+}]$ in lake water balanced the [Alk] (effectively HCO_3^-), that no weak acids were present, that other alkalinity was absent, and that sea salts can be ignored. The chemistry of 13 lakes in areas 'unaffected by acid rain' (aside from a 'world average' value, these were from Scandinavia and North America) was used to provide a universal 'pre-acidification alkalinity' regression of bicarbonate against the sum of calcium and magnesium:

$$Alk_0 = 0.93(Ca^* + Mg^*) - 14$$

where Ca^* and Mg^* are non-marine concentrations. The deviation of present-day alkalinity (Alk_t) from (Alk_0) was referred to as the **acidification index, AI**:

$$AI = Alk_0 - Alk_t$$

To maintain electroneutrality when sulphate is increased, a decrease in other anions or an increase in cations is required. The early Henriksen model assumed that calcium and magnesium concentrations are constant for any catchment, but a later modification introduced a factor F, representing the ratio of changes in these cations to changes in sulphate, implying greater cation mobilization as sulphate loading increased. The equation was then modified to predict alkalinity (Alk_z) from changes in base cations and sulphate:

$$Alk_z = 0.93(Ca^* + Mg^*) - 14 - F(SO_4^{2-*})$$

Values for F were found to vary from 9 to 0.4 for Norwegian lakes (Henriksen, 1982).

The application of the Henriksen index to 94 UK data sets for which sufficient data for alkalinity, pH and base cations for a five-year period since 1980 were available proved disappointing (Warren *et al.*, 1988). Assuming that the model takes account of the bulk of the ionic chemistry, the index, AI, should be positively correlated with non-marine sulphate; a strong relationship was indeed found between this and alkalinity with a few anomalous (possibly erroneous) data points. But the high degree of seasonal variation requires that the data should be representative of year-round conditions, and differences for sites varying in land use (moorland versus forested) and marine influence, as well as the simplified chemistry, limit realistic applications of the AI concept, at least in UK conditions. Further, short-term changes in rain or surface water acidity are not included, although episodic events may be the most damaging. Finally, the AI values derived did not provide a synoptic picture of the degree of acidification in the UK, and high values did not necessarily indicate greater acidification or the scale of biological response.

The recent NAPAP exercise considered the relationships of anion and cation species in lakes and streams, although not quite along the lines of the Henriksen model (L. A. Baker *et al.*, 1990). For the northeast data set as a whole, the change in the ratio of ANC and sulphate and nitrate, and of calcium and magnesium against these acid anions, was -0.13 and $+0.54$ respectively. Somewhat different values were derived from palaeolimnological records. This led to estimation of the loss of

ANC of 18 μeq l^{-1} for the Adirondack lakes over 150 years, 7 μeq l^{-1} for other lakes of low current ANC levels, and none for those of current ANC > 25 μeq l^{-1}. The analysis also included changes in inorganic aluminium which increased with the change in acid anions.

A review of various steady-state models for projecting pH changes (Thornton *et al.*, 1990) concluded that the empirical models based on manipulations or simplifications of the charge balance model (e.g. Almer *et al.*, 1978; Henriksen, 1980; Wright and Henriksen, 1983) were not acceptable for the NAPAP integrated assessment, because they could not predict conditions over time, nor take account of catchment or biological response outside the regions for which they were developed. The more dynamic catchment models (MAGIC, ILWAS) were favoured (see Chapter 11); these take account of the complex but important soil/water transfers of ions.

6.5   PERMANENT OR TRANSIENT ACIDITY

Surface waters are sometimes characterized as being *permanently* acid. However, many lakes and most streams are subject to seasonal or more rapid changes in pH, alkalinity or other relevant chemistry, usually during hydrological events over timescales of hours to weeks. This instability is particularly apparent in waters of low conductivity or alkalinity, often upland headwaters draining catchments of low base reserves. Sudden rainfalls lead to rapid runoff, with little contact time with catchment soils. These conditions provide susceptibility to acidification, even in areas receiving virtually background levels of acid rain, for instance at sites in northern Sweden (Bjarnborg, 1983; Townsend *et al.*, 1990). This almost ubiquitous phenomenon has some exceptions, with a few chronically acid waters not experiencing such episodes (Wiginton *et al.*, 1990). Change in the hydrological flowpath of drainage water is the most important factor determining the scale and duration of acid episodes. This arises because the more rapid runoff directs drainage waters through or over upper, more acid soils, rather than allowing contact with the less acid mineral lower horizons. Soil processes such as nitrification, organic acid production and oxidation during drought conditions are all involved. Man-made acidic components in deposition also contribute to episodes if they have accumulated in a catchment during dry periods, to be flushed out by heavy rain. Upwind events may also contaminate rain or snow with effects on snow and meltwater at distant sites, as after an oil fire off Holland in 1987 (Tranter *et al.*, 1987, 1988).

Some of the earliest records of acid episodes in freshwater systems were seen during snowmelt or following heavy rain after a prolonged dry spell (Sullivan *et al.*, 1986; Fig. 6.7). In the case of snowmelt, fractionation within the snow pack leads to concentration of the ionic materials present and depression of the freezing point. As a result, the first meltwater may have concentrations about five times greater than that of the bulk snow; 50–80% of the accumulated ions are mobilized in the first 30% of meltwater (Seip *et al.*, 1980).

Acid stream events are also observed in association with 'sea salt' rains in maritime areas, where the high concentration of sodium and chloride can displace H$^+$ and aluminium held on the soil exchanger, generating highly acidic flushes, even though

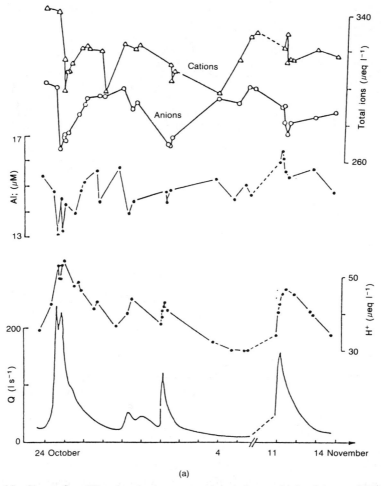

(a)

Fig. 6.7—Stream flow (Q) and major ion concentrations, inorganic aluminium and H^+ at Birkenes, southern Norway: (a) after a protracted dry spell in autumn 1984; (b) after snowmelt in spring (after Sullivan *et al.*, 1986).

rain pH is quite high. Such episodes recorded in Galloway, southwest Scotland (see Chapter 3, Fig. 3.6; Langan, 1987, 1989) may raise acidity from 3 to 25 μeq l^{-1} over a few hours, an acid flux two- to ninefold greater than that delivered in the rain. This phenomenon is also typical of sites where soils are acid and peaty where a sudden rainfall can flush out water of low ANC held within bogs and peat formations (Kleissen *et al.*, 1990; Wheater *et al.*, 1990).

While acidity increases as ANC falls during episodes, the other chemical species behave in ways specific to the site conditions. In Norway, snow or rain episodes have reduced pH by as much as 1 pH unit, doubling sulphate and nitrate concentrations and increasing aluminium as much as tenfold. Elsewhere in Norway snowmelt led

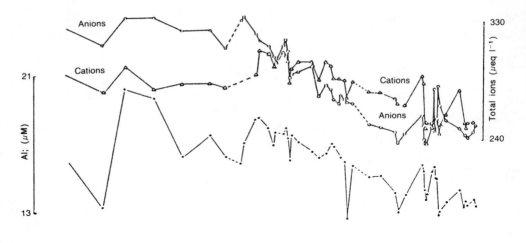

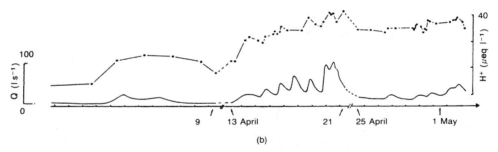

Fig. 6.7—*continued.*

to increases in both acidity and sulphate of rivers (Johannessen *et al.*, 1980). Even in a pristine brook with carbonate buffering and weak organic acidity, snowmelt pH fell from 7.2 to 5.0 at high flow, even to 4.8 (Christophersen *et al.*, 1990), in this case attributed to sea salts. In Pennsylvania, USA, sulphate pulses result in significant ANC depressions, while in the Adirondacks and Catskills in New York State nitrate is strongly associated with episodes, indeed sulphate sometimes falls (Fig. 6.8; Wiginton *et al.*, 1990). A similar variety of response, even at the same site, is documented for Sweden, the UK and Canada, limiting generalizations.

Further evidence of the nature of episodes has been obtained from field experiments. In one (Henriksen *et al.*, 1988), dilute sulphuric acid was added in excess of the alkalinity present by 20-fold, reducing the pH from > 6.0 to about 4.0—a factor of 100 in [H^+]. The pH drop was accompanied by an increase in labile (inorganic) aluminium. Some increases in other cations were seen, equivalent in total to the added sulphate, and small amounts of trace elements were also released. The episode took place over nine hours, and recovery time was similar. In another experiment

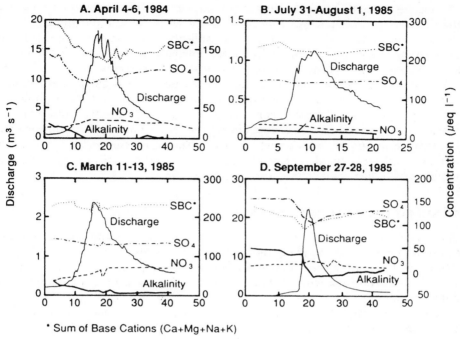

* Sum of Base Cations (Ca+Mg+Na+K)

Storm Duration (hours)

Fig. 6.8—Chemical changes during episodes of high flow at Biscuit Brook, Catskills, New York (after Murdoch, 1988).

(Abrahamsen *et al.*, 1979), an artificial rain was applied to an expanse of almost vegetation-free granite bedrock with runoff similar to that of the incident rain. Use of a salt-free 'rain' led to a rise in pH as the H^+ exchanged with metal cations.

An increase in aluminium during episodes is consistently observed, mobilized by more acid conditions in soils. Some, however, is thought to be due to release from sediments and vegetation within the stream (Tipping and Hopwood, 1988) and a field acidification of a Welsh stream (Ormerod *et al.*, 1987) from pH 7.0 to 4.3 increased aluminium sevenfold.

6.6 UNRESOLVED ISSUES

Although there is now a large compendium of data world-wide regarding the chemistry of acid waters, there are still some aspects that are unresolved.

Relationships between deposition and surface water composition over space and time are particularly important with regard to the prediction of water quality changes in response to emission changes.

The contribution of weak acids is not well understood; some observations (e.g. RAIN) suggest that as strong acids fall, more acidity is associated with weak acids.

An increasing contribution of nitrate is reported, but the extent to which this is determined by atmospheric input or in-soil processes is not well quantified.

Specific site differences are not always explicable from current knowledge of catchment soils and vegetation, hydrological flowpaths and atmospheric inputs to the catchment. This is particularly true of short-term acid events.

The contribution of reservoirs of sulphur in catchment soils is quite uncertain; their transfers to surface water and their transformations within the soil are important in predicting the benefit of emission controls to surface waters and their timescale.

Prediction of the response to changes in atmospheric input over both the long and short term needs further development and generalization for characteristic catchment conditions. Dynamic models (e.g. MAGIC) incorporating soil and hydrological conditions have had some limited success in this respect.

6.7   SUMMARY AND CONCLUSIONS

Deposition plays a major role in the chemistry of surface waters, and acid components are also important in determining the composition of surface waters.

However, the pattern and volume of rainfall is as important as its quality, and interaction of deposition with vegetation and soils is critical to the development of acidified conditions.

There is no universal consensus as to what is an acid or acidified water—pH values from 6.5 to < 5.0, or ANC limits of 50–0 μeq l^{-1}, are quoted according to interest or conviction; empirical ionic ratios or anion/cation balances have been used to characterize ANC.

In any event, the chemistry of such unbuffered oligotrophic waters is highly dynamic; any characterization should be based on long-term, flow-weighted, data sets, or a consistent sampling pattern.

Sulphate is the major anionic constituent strongly linked with acidification but over the last 15 years nitrate has increased, and sometimes even exceeds the contribution from sulphate.

Attempts to develop an empirical 'acidification index' from ionic relationships have not been helpful for understanding mechanisms of acidification or prediction of future changes.

Organic acids generated by vegetation and soil processes are significant in generating soil anions, and drainage from acid soils, peats or bogs can reduce runoff water to pH < 5.0; in some cases runoff is more acid than the incident rain. At pH < 5.0, strong mineral acids become more important.

The cationic species in surface waters are derived from the atmospheric input and from weathering processes in soils. The neutralizing base cations, calcium and magnesium, are mostly derived from catchment soils and are dependent on mineral materials and sufficiently acid conditions to release them.

The deposition of sea salt material to peaty acid soils of low base reserve displaces H^+ and Al^{3+} from the soil exchanger and leads to acid episodes in streams.

7

Spatial and temporal trends

7.1 INTRODUCTION

Although most of the world's waters are not acid, the popular concept is that surface waters receiving acid rain (defined as pH $\leq$ 4.7, Wright and Henriksen, 1979) will be acid (pH $\leq$ 5.0); acidification of sensitive waters has occurred over the past 100 years. In Chapters 5 and 6 these criteria are examined critically. The regional extent of acid water distribution, the degree of sensitivity and the rate of acidification are often uncertain; this chapter will review some synoptic surveys of acidified lakes and streams.

Over the past 25 years a number of extensive surveys has been undertaken and in some cases sites have been revisited over a passage of time to document temporal changes. In addition an extensive number of lakes has provided evidence of longer temporal change on the basis of diatom remnants or other fossil material in lake sediments. In few instances has the sampling followed a statistical protocol allowing assessment of the distribution of acid waters within a lake or stream population for a region. In many cases, lakes or streams have been selected because of their identified acidity, providing an impression that they are representative of a region or country, whereas adjacent waters may not be acidified, or may be to a lesser degree. Such observations should stimulate an investigator to explore this difference, rather than to obscure it by generalization.

In Europe, acid-sensitive areas have been identified not only in Scandinavia, but also in sensitive parts of most western European countries. In North America, northeastern states and bordering provinces of Canada lying on the Canadian Shield, as well as the maritime areas of Nova Scotia and Labrador, have a substantial number of acid waters; they are also found in Florida and in some areas in the northwest, in the southern Rockies. Further afield, streams in the Amazon basin, dune lakes in Australia, and waters in southwestern New Zealand are also acid. This diverse group includes seepage as well as drainage lakes, and some with high humic content; the importance of acid rain as the major source of acidity clearly varies with hydrology, catchment soils and vegetation as well as with the contribution of acidic

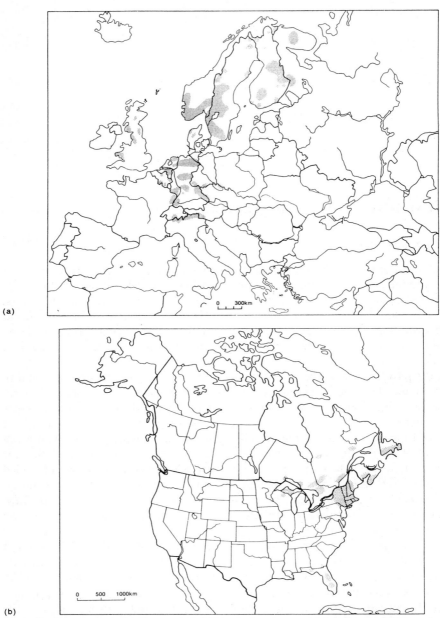

(a)

(b)

Fig. 7.1—Areas in (a) Europe and (b) North America reported to have acidified surface waters.

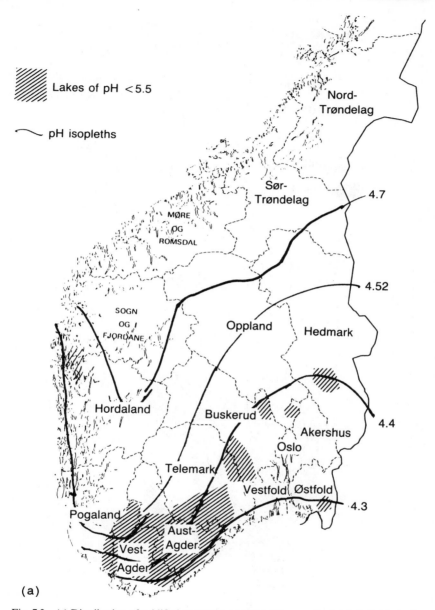

Lakes of pH <5.5

pH isopleths

Nord-
Trøndelag

Sør-
Trøndelag   4.7

MØRE
OG
ROMSDAL

4.52

SOGN
OG
FJORDANE   Oppland   Hedmark

Hordaland   Buskerud   4.4

Akershus

Oslo

Telemark

Vestfold Østfold

Pogaland   4.3

Aust-
Agder

Vest-
Agder

(a)

Fig. 7.2—(a) Distribution of acidified waters in southern Norway; the isopleth lines are the
rain pH. (b) Distribution of acidified waters in Sweden; the isopleths indicate levels of
sulphate, based on 71 reference lakes.

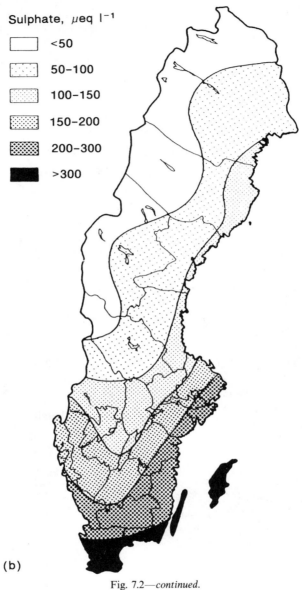

Fig. 7.2—*continued.*

materials in deposition. Principal European and North American areas with identified acidified waters are shown in Figs 7.1 and 7.2; it is these areas which have been most intensively studied. The requirement that they received acid rain of pH $\leqslant 4.7$ is clearly not always satisfied.

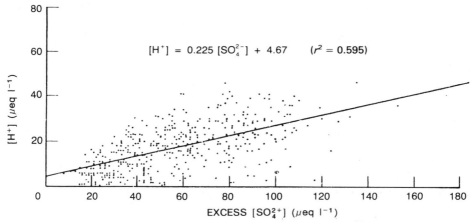

Fig. 7.3—The relationship of non-marine sulphate and acidity in 471 southern Norwegian lakes; acidity [H^+] reflects about one-fifth of lake [$SO_4{}^{2-}$] (after Brown and Sadler, 1981).

7.2 DISTRIBUTION AND FREQUENCY: SOME REGIONAL SURVEYS

Large regional estimates are based on samples of lakes and rivers, often large, taken from across a broad area. Inevitably they are 'spot' samples, sometimes specified by time, e.g. prior to snowmelt, or at post-summer turnover. Analyses are often limited in their scope and depth, with inconsistent inclusion of some parameters such as alkalinity. It is not always certain that measurements are accurate or consistent, nor that the estimation methods were valid (L. A. Baker *et al.*, 1990).

Norway

In southern Norway, an early regional survey of 128 lakes (Wright, 1977) provided evidence of many acid waters in this region. Two lakes (> 20 ha) were selected in each 10 km^2 area to give a geographical perspective. Later surveys (700 lakes) in the same region (Wright and Snekvik, 1978; Henriksen *et al.*, 1987) showed that the distribution of waters of pH $\leqslant$ 5.0 was concentrated in the southwest where the pH of rain was $\leqslant$ 4.3. It was reported that in this area 'thousands of lakes and streams...have ANC < 0'. The recent '1000 Lakes' survey (Fig. 7.2(a); Henriksen *et al.*, 1987) has shown that lakes in southern Norway are very low in base cations— < 75 μeq l^{-1} in 75%, and even < 25 μeq l^{-1} in 50%; the lakes were not selected statistically. In the recent survey, 40% had pH < 5.0, 60% < 5.5, mostly in the south, although a few were found in eastern and western Norway. The most acid lakes have the highest sulphate concentrations, as well as the lowest concentrations of basic cations. Non-marine sulphate concentrations are also greatest in the south and southeast, matching the pattern of sulphate deposition. Analysing the synoptic data in nearly 500 lakes, Brown and Sadler (1981) found a reasonable relationship between lake acidity and excess sulphate (Fig. 7.3) although only about a fifth of the sulphate is matched by acidity. In general, organic materials are low (TOC < 6 mg l^{-1}) and organic anions account for < 10% of all anions. Nitrate concentrations were also highest in the south, and increased significantly in the period from 1974 to 1986 in the resampled lakes, whereas sulphate decreased slightly; nitrate is now second only to sulphate in concentration.

Sweden

In Sweden, of an estimated total of 85 000 lakes, 4000 lakes are classed as 'seriously acidified', i.e. reported on some occasion to have pH $\leqslant$ 5.0, with negligible alkalinity, on the basis of samples taken from nearly 7000 lakes. Further, at least 18 000 are thought to be transiently acid during critical periods (high flows, snowmelt) (MONITOR, 1985). Predominantly, the acid lakes are small, usually < 10 ha in area (87% of those of pH < 5.0). Sulphate was not measured in the full sample of lakes, but it was analysed in 170 monitored special interest sites. Most lakes of pH < 4.9 were found in the southern or southwest part of the country (Fig. 7.2(b)), and concentrations of sulphate were highest there (300 μeq $SO_4^{2-} * 1^{-1}$); they were lowest in the north (< 50 μeq 1^{-1}). The distribution of acid waters reflects the pattern of sulphate deposition and is considered evidence that 15 000 lakes were acidified because of air (S) pollution; an additional 6500 were acidic due to organic, humic materials (Bernes and Thornelof, in L. A. Baker *et al.*, 1990). The data from the 1985 survey are incomplete and some of the measurements judged inconsistent (L. A. Baker *et al.*, 1990). In addition to sulphate, nitrate was also high (> 10 μeq 1^{-1}) in many acid lakes.

Finland

Much of Finland is considered sensitive to acidification with thin podzolic soils lying over granitic bedrock. There are 56 000 lakes > 1 ha and about 15 700 > 10 ha; nearly 1200 were sampled in 1987 with a predominance of small lakes in the southern region. The selection of lakes was random but statistical design was such that sampling was considered representative (Kamari *et al.*, 1991; L. A. Baker *et al.*, 1990). Median pH was 6.1, but 4100 lakes (10%) had pH $\leqslant$ 5.0 and more than 10% had ANC $\leqslant$ 0, while 18% had ANC < 10 μeq 1^{-1}. The lakes are high in humic material (91% with TOC > 5 mg 1^{-1}) which is an important acid component; indeed organic anions exceeded sulphate. In central and southern Finland, 44% of lakes had TOC $\geqslant$ 15 mg 1^{-1}, especially those draining peat. Even so, the geographical pattern of lake sulphate was similar to that of deposition (Kortelainen and Mannio, 1988), with highest concentrations (100 μeq 1^{-1}) in southern Finland. Base cations (Ca + Mg) were higher than those in Norway (median concentration 140 μeq 1^{-1} in the south) but about 10% had < 100 μeq 1^{-1}. Nitrate levels were low (only 1.1 μeq 1^{-1}). Empirical ion relationships indicate that 74% to 87% of Finnish acid lakes are naturally acidic (L. A. Baker *et al.*, 1990); compared with areas such as Norway, most Finnish lakes are not considered particularly sensitive (Kamari *et al.*, 1991).

United Kingdom

Areas in Britain considered susceptible to acidification are identified in relation to geology, soil type, land use and rainfall—they are predominantly in western and upland areas (Fig. 7.4; Patrick *et al.*, 1991). However, data on pH and alkalinity over a 10-year period provided by water authorities for 75 national river monitoring sites in relevant areas throughout England and Wales provided little evidence of acidified waters or trends of increased acidity (Fig. 7.5). For 18 Scottish sites there was a fall in pH from 6.6 to 6.3 between the late 1960s and 1983–85, but this change was unrelated to changes in sulphate (Warren *et al.*, 1986). However, it was accepted that these waters did not represent the small upland tarns and streams more likely to be affected. Where these have been investigated (though not on a regional basis) it is clear that they are in areas of high rainfall and base-poor bedrock and thin soils rather than in areas of low rain pH or high sulphate concentration, although they have high deposition; chloride is the major anion present. Streams and lakes receiving runoff from afforested areas are more acid (with higher sulphate and aluminium concentrations) than those in the same area draining moorland (Warren *et al.*, 1988).

Rest of Europe
Many other surveys are now reported for other countries in Europe (Nordic Council of Ministers, 1988)—Denmark, northern Germany, the Netherlands, Austria, Switzerland and Italy (Fig. 7.1(a))—usually in areas of sandy soils, hard bedrock and afforested terrain. In general, the choice of the sampled waters has not been systematic or planned statistically to represent the lake population. A notable feature in common is the great variation between lakes, even when they receive comparable deposition and drain catchments with similar soils and vegetation. The explanation of this diversity is presumably that hydrological conditions play a major role.

United States (Fig. 7.1(b))
The NAPAP and associated lake/stream sampling programmes cover most of the eastern states, with some samples also in the west and upper midwest. In all, 1592 lakes of > 4 ha, representing a population of 17 953 lakes, were sampled in the eastern survey and 719 lakes of > 1 ha of a 10 393 population in the western survey. In addition, 500 streams representing 56 000 stream reaches were sampled.

The majority of lakes and streams are circumneutral (pH > 6.0) and weakly buffered (half with ANC $\geqslant 200 \ \mu\text{eq} \, l^{-1}$); an estimated 4.2% of lakes are acidic, defined as ANC $\leqslant 0 \ \mu\text{eq} \, l^{-1}$. About a third of the acid lakes are in Florida, and about a third in the northeast, of which 181 are in the Adirondack region. There are virtually no acidic lakes in the interior southeast or western regions. Acidic streams in the Appalachians and Catskills (excluding those subject to mine drainage) usually drain catchments of $< 30 \ \text{km}^2$ and are at altitude > 300 m; they are usually low in DOC and high in aluminium. Further south, streams have low gradients and exhibit large variations; some are high in DOC, and inorganic aluminium is highly variable.

Inorganic acidity dominates acid lakes in the northeast and Florida, with sulphate the dominant anion (median 102 $\mu\text{eq} \, l^{-1}$); nitrate is low ($< 1 \ \mu\text{eq} \, l^{-1}$) and DOC is also low (2 mg l^{-1}), reflecting a low occurrence of wetlands. For the sampled streams, most (70%) are similarly dominated by sulphate acidity, while about 30% are organic acid waters; a few (33) are dominated by catchment sources of sulphur. Mine drainage is a major cause of acidic streams in much of the northeastern region; these were

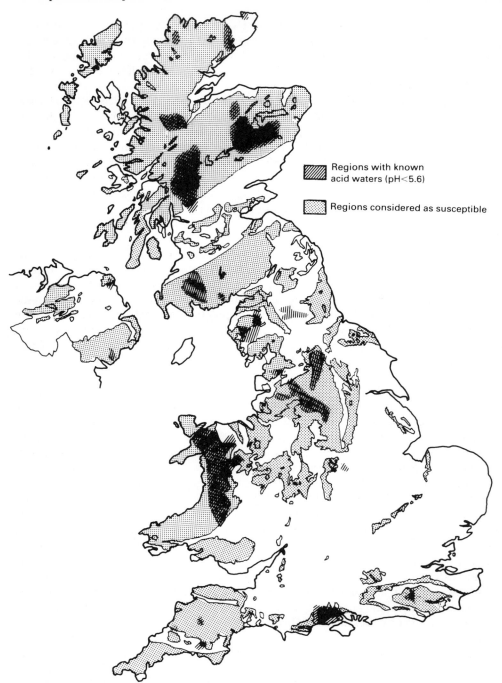

Regions with known acid waters (pH<5.6)

Regions considered as susceptible

Fig. 7.4—Areas in the UK sensitive to acidification (stippled) and where acid waters are reported (cross-hatched) (after Warren *et al.*, 1988).

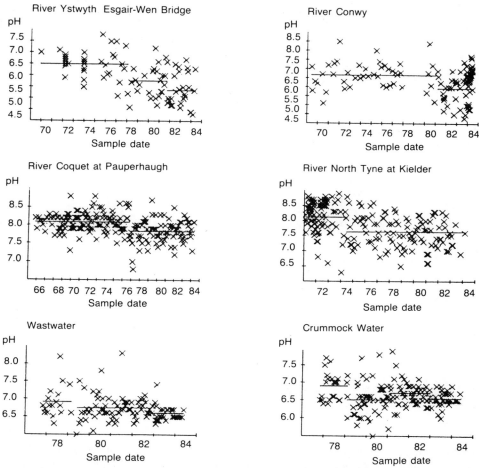

Fig. 7.5—Acidification trends in recent decades in six rivers in susceptible areas of the UK
(after Warren *et al.*, 1986).

excluded from analysis since the objective was to characterize those waters affected by acid rain.

While a relationship was found between rain sulphate deposition and lake and stream sulphate concentrations, such a linear relationship was not found with ANC or pH, because of the intervention of catchment influences. However, all lakes and streams with pH $\leqslant 5.5$ or ANC $\leqslant 0$ are in areas of rain of pH $\leqslant 5.0$ and wet sulphate deposition of > 10 kg ha^{-1} yr^{-1}. In contrast, nitrogen deposition $(NH_4^+ + NO_3^-)$ was not so simply related to lake nitrate concentrations, although to some extent this was true of streams. Most of the acidic waters sampled fall into several 'high interest' regions; it was conceded that earlier focus on these regions led to an overestimate of the extent to which US waters had become acidified (Irving, 1991).

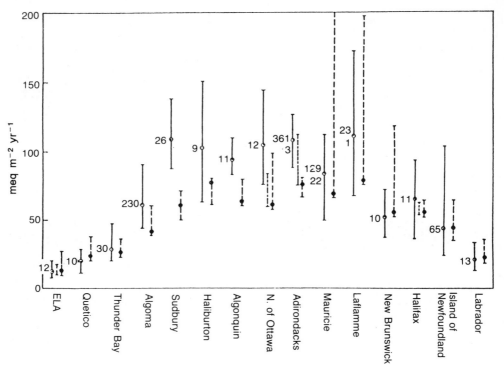

Fig. 7.6—Transect of North American regions from Labrador to western Ontario, showing non-marine sulphate deposition (——) and catchment yields of sulphate (——) (1980 data). The numbers refer to the sites sampled.

Canada

In southeastern Canada (about 16% of the national area) where there are nearly 400 000 lakes, 8506 have been sampled and analysed. The acid waters lie in Ontario, Quebec, Labrador, Newfoundland and Nova Scotia. The spatial coverage of sampled lakes varies between provinces (Fig. 7.6) and suggests that both non-marine S deposition and sulphate yield are highest in Ontario. Of the large data set, 173 sampled lakes were excluded because of missing or inconsistent variables, and a larger group (4316) were omitted as they were not considered sensitive to acid deposition for a variety of reasons. The remainder (4017) is sufficient for a detailed examination of acid–base chemistry on a regional scale.

About 56% of the lakes have ANC $\leqslant 100 \ \mu$eq l^{-1}, most in southeast Ontario (69% of lakes there); about 38% of all sampled lakes had ANC $\leqslant 50 \ \mu$eq l^{-1} but only about 5% (381) had zero ANC. Few lakes elsewhere in Ontario are acid, and if so are characterized by high DOC levels. There is a significant correlation of lake water sulphate with sulphate deposition, as for the US data set, supporting this as the principal source. However, the lack of a strong relationship of sulphate and ANC (or Ca + Mg) suggests that bases are controlled by catchment characteristics, even for lakes of ANC $\leqslant 50 \ \mu$eq l^{-1}. Organic anions are estimated as 25–60 μeq l^{-1}

and, along with HCO_3^-, add to the acidity. For lakes with zero ANC and $[SO_4^{2-}] > [Ca^{2+} + Mg^{2+}]$ about half (106 of 230) have low organic anions and their acidification is attributed to sulphate deposition alone; they are located primarily in the vicinity of the Sudbury smelters. In Nova Scotia, lakes of zero ANC have organic anions in excess of sulphate, and acidity is attributed to organic acids. For the remainder, sulphate acidity and organic acidity together have led to acidification.

Rest of the world
Streams of the Amazon basin are typically acidic; many are coloured and humic acids evidently contribute to their acidity. Some are identified as **clearwater** systems of pH 4.5 to 4.9 (Junk, 1983) but it is not clear whether their acidity is due only to strong inorganic acids. The waters are certainly of uniquely low ionic strength (Stallard and Emond, 1983), thus low concentrations of base cations can be assumed.

Dune lakes are found along the eastern coast of Australia where wind erosion leads to the formation of a perched water table which seeps to the lakes. They are clear or coloured, the latter seeping from humic podzols (Reeve and Fergus, 1982). Their chemistry is strongly influenced by sea salts, with sodium and chloride the dominant ions; other base cation concentrations are low. Acidity in many is attributed to organic acids, even at pH < 5.0. Elsewhere (e.g. Frazier Island), clearwater acid lakes are found; they are low in sulphate and their acidity has been attributed to base cation retention (Krug, 1989). Rain in Sydney is reported as having pH $\leqslant 4.7$ and non-marine sulphate is 24 μeq$\,l^{-1}$, so acidic deposition cannot be ruled out as the cause of acidification. Some lakes in Tasmania also have pH $\leqslant 5.0$, apparently due to organic acids; pH was correlated with the degree of colour and the estimated DOC of acidic lakes was as high as 20 mg$\,l^{-1}$.

Lakes in the southwest of New Zealand are often acidic, as low as pH 3.3, evidently affected by high humic concentrations (estimated DOC 8–16 mg$\,l^{-1}$) (Collier and Winterbourne, 1987; Krug, 1989). Very acidic lakes also occur in areas of volcanic activity (McColl, 1975), either acid–sulphate–chloride waters formed when neutral chloride waters dissolve mineral sulphides, or acid–sulphate waters where steam carrying H_2S condenses and then oxidizes. These waters have very high sulphate concentrations (> 300 μeq$\,l^{-1}$) and very low pH, to < 1.8.

7.3   SOME INTENSIVE STUDIES

Surveys provide information about the extent and degree of the current acid status of surface waters, but seldom help the assessment of causes or timescales of change, even where correlations are found.   More insight comes from more intensive studies of a few lakes or streams where sampling has continued over extensive periods and where additional measurements are available regarding deposition, soils, vegetation and flows, so that relevant processes, mechanisms and relationships can be illustrated. Some examples are given here.

English Lake District

Data for lakes and tarns (ponds) in Cumbria from more than 35 years of sampling, often using the same methods of analysis, have been collated and reported (Sutcliffe *et al.*, 1982; Sutcliffe, 1983; Sutcliffe and Carrick, 1973, 1988). In this region of heavy rainfall, sulphate concentrations in 1988 were about $60\ \mu eq\ l^{-1}$ and H^+ about $27\ \mu eq\ l^{-1}$ (annual wet deposit $> 0.05\ g\ H^+\ m^{-2}$, $> 1.6\ g\ S\ m^{-2}$; Irwin *et al.*, 1990); non-marine sulphate is about 80%. About half of the lakes and tarns (31 of 60 sampled) and many streams draining igneous rocks are permanently acidic and oligotrophic, mostly those in the western, central and northern fells. These upland waters are low in base cations (about $30\ \mu eq\ l^{-1}$) and alkalinity, and also in non-marine sulphate (about $40\ \mu eq\ l^{-1}$), but are sensitive to acidification. Lower-altitude lakes draining Silurian slates in the southern region have higher alkalinity and higher concentrations of chloride and sulphate, as well as nitrate; they are not so sensitive and are quite productive. Of 75 tarns sampled in this area, only six had summer alkalinity values $< 100\ \mu eq\ l^{-1}$. In this diverse geological area, adjacent lakes may show quite different chemistries reflecting their geology and hydrology rather than differences in deposition. Over a 30-year period, few tarns have changed in their acid status; four, all draining an area of Silurian igneous rocks, have become more acid since about 1900.

In the same region, streams of the upper Duddon catchment have been sampled repeatedly. Comparing the chemical data of the 1970s (Carrick and Sutcliffe, 1983) with data for 1986 (Tipping, 1989), it appears that sulphate concentrations declined from about $116\ \mu eq\ l^{-1}$ to about $75\ \mu eq\ l^{-1}$, a fall of about 40%. In contrast, nitrate concentrations have increased from about 10 to $30\ \mu eq\ l^{-1}$. While the Duddon catchment was apparently a net sink for nitrogen in the 1970s, by 1986 it was neither a source nor a sink. Production of H^+ was about $13\ \mu eq\ l^{-1}$ in the earlier period, increasing to about $59\ \mu eq\ l^{-1}$ in the 1980s. However, acidity in the streams seems to have changed little, along with most major ions, and it is concluded that the change during the decade could be attributed to enhanced deposition of ammonium, oxidizing to nitrate within the catchment. In this study, stream concentrations are determined by deposition input but are also affected by evapotranspiration, plant N uptake, dissolution of $Al(OH)_3$ from soil, precipitation of $Al(OH)_3$ to the stream bed and carbonate reactions. Organic carbon is low, below $1\ mg\ l^{-1}$, along with low aluminium, neither contributing to the acidity/alkalinity balance (Tipping, 1989). Some sulphate retention in the catchment was acknowledged, but this is likely to influence rates of change rather than steady-state concentrations.

Southwest Scotland

Loch Dee and Loch Fleet in Galloway have been sampled intensively over a decade or more (Lees *et al.*, 1989; Lees and Farley, 1993; Welsh and Burns, 1987; Howells and Dalziel, 1992). Like other lakes in this area (Fig. 7.7) they receive runoff from terrain underlain by granites, with glacial tills and peaty or podzolized soils above; conifer forest covers $< 1\%$ in the Dee catchment, $< 10\%$ at Fleet. Dee has a significantly greater base weathering capacity (Langan and Harriman, 1993) than Fleet (*ca.* $< 9\ meq\ m^{-2}\ yr^{-1}$; Skeffington and Brown, 1992). Rainfall is heavy

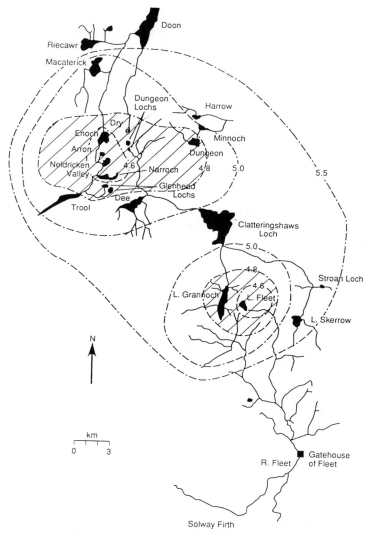

Fig. 7.7—Acid lakes centred on the granite plutons (hatched) in Galloway, southwest Scotland; the isopleths show the limits of lake acidity.

(2200 mm yr^{-1}) with pH 4.7–5.0 and sulphate about 50 μeq l^{-1}; sulphate deposition is high, 5.2 g and 6.0 g SO$_4^{2-}$ m^{-2} yr^{-1} respectively. Non-marine sulphate is about 80% of the total. Rain delivers a large sea salt component, and as a result chloride is the major anion in the lakes; episodic 'sea salt rains' lead to high acid and aluminium pulses in runoff (Chapter 6). Loch Dee pH is about 5.6, that of Fleet (preliming) was 4.0 to 4.5; calcium is low, $\leqslant$ 50 μeq l^{-1}, and labile aluminium is high. The lakes are oligotrophic, especially Fleet where phosphorus is non-detectable and primary production low.

Table 7.1—Flow weighted chemistry (μeq l^{-1}) of an acid moorland and an acid forest stream in the Llyn Brianne catchment, Wales

Site	pH	SO$_4$	Cl	Na	Ca	Mg	Al^t
Moorland stream	5.2	90	169	146	54	62	170
Forest	4.6	134	219	172	52	55	520

Bulk rain sulphate influx underestimates total deposit of sulphate, especially in runoff below forest, but effluxes exceed influx at 9.2 g SO$_4{}^{2-}$ m^{-2} yr^{-1} at one Dee sub-catchment without forest (Dargall Lane) and at 11.3 g m^{-2} yr^{-1} at Fleet outflow, suggesting that catchment sources of sulphur (including accumulated marine sulphur) may contribute.

There is little evidence of a consistent trend in either pH and sulphate of rain or lake chemistry at either Dee or Fleet through a decade of intensive sampling, from 1980 to 1990 (Lees, 1992; Howells *et al.*, 1991).

West Wales: Llyn Brianne
At this base-poor site in Wales, water chemistry data for 14 streams illustrate differences attributable to land use (Bird *et al.*, 1990; Donald and Gee, 1993). Six moorland streams are included, one with circumneutral pH (6.82), relatively high hardness (14.5 mg l^{-1} as CaCO$_3$) and low aluminium (50 μg l^{-1}), the others with lower pHs and hardness but higher aluminium; sulphate, nitrate and chloride levels are broadly similar. Winter conditions are characterized by lower pH and calcium levels, as well as higher nitrate, sulphate and sodium; acidity, nitrate and aluminium are significantly correlated with flow (log). Within the same catchment, five streams drain conifer-afforested areas. They are notably lower in pH throughout the year even with similar or higher hardness categories, and significantly higher aluminium (Table 7.1). As with the moorland streams, pH and aluminium are correlated with flow and with calcium. Another stream draining an oak forest is less acid, and aluminium levels are low, although sulphate is higher than in moorland streams; seasonal differences were not striking there and flow correlations were poor.

Southwest Sweden: Gardsjon
Streams and lakes in the largely forested Gardsjon area have been subject to detailed chemical analysis for nearly two decades (Andersson and Olsson, 1985). The lakes have been considered representative of many other surface waters in southwest Sweden sensitive to acidification (Hultberg, 1985). Rain in the area has SO$_4{}^{2-}$ 42 μeq l^{-1}, hydrogen 20 μeq l^{-1}; annual deposit is 3.4 g m^{-2} for sulphate, 0.05 g m^{-2} for hydrogen (Hultberg, 1985; Nilsson, 1985; Hornung *et al.*, 1990); about 86% of sulphate is non-marine. The lakes have a range of pH values from 4.68 to 5.28, low base cation concentrations (Ca^{2+} + Mg^{2+} are 55–115 μeq l^{-1}), high sulphate (to 250 μeq l^{-1}) and low nitrate or ammonium; some major ions (Mg, Na, K) are derived from weathering while others (Na, Cl) are from sea salts. Aluminium, manganese and iron are present but insignificant in terms of the charge balance;

similarly, organic ions (DOC 2–5 mg l^{-1}) are thought unimportant. Major cations are balanced by the anions chloride, sulphate and nitrate. Seasonal changes reflect changes in runoff and vegetation through the year, as well as differences in rainfall. For Lake Gardsjon, a proton budget suggests that about half the proton input is attributable to atmospheric deposition, and about half to the protons and aluminium generated in soils, the latter possibly enhanced by the high sulphate flux and accompanying soil acidification (Nilsson, 1985).

The intensive study at Gardsjon has allowed the calculation of input/output flux budgets for acidity and sulphate (and other ions). For all protons, expressed as H$^+$, annual input is estimated to total about 0.3 g m^{-2} (or 3 keq ha^{-1}), while proton consumption was equivalent, 2.7–3.6 keq ha^{-1}, suggesting that there is no progressive acidification (Nilsson, 1985). However, the lake efflux accounts for a loss of 0.94 keq H$^+$ ha^{-1} and the input/output ratio is 1.17–1.58, an annual excess of about 0.5 keq, implying that the balance is neutralized within the lake. For sulphate, total annual influx to the lake is virtually matched by the efflux, taking account of lake volume and some loss to sediments (Hultberg, 1985). However, on the basis of bulk deposition input, lake efflux shows an almost threefold increase during transfer through the catchment (Hornung et al., 1990), attributed to enhanced catchment scavenging and release of soil sulphur reserves.

In another independent study of 20 small acidified lakes in the same region (including the Gardsjon lakes) (Morling, 1981), data gathered between 1966 and 1979 showed a significant increase in acidity by as much as 2 pH units since the 1940s, mostly in the decade 1965–75. Alkalinity in the lakes fell to virtually zero. Sulphate concentrations increased about threefold, while chloride also increased by 50%. While some of these changes were attributed to an increased influx of sulphate and nitrate from distant sources, changes in climate were noted during this period, with lower rainfall and reduced levels of groundwater, as well as a significant sea salt input, making interpretation uncertain; changes in land use in the area (deforestation) also occurred during the study period. Nonetheless it was judged that changes in lake sulphate closely followed those in sulphate deposition; both increased by about 1% per year.

Southern Norway: Birkenes

A stream in southern Norway (Birkenes I) has been sampled intensively for 20 years with detailed chemical analysis, providing data for modelling the acidification process (Christophersen et al., 1990). The site has the highest annual deposition of sulphate (6 g SO$_4$$^{2-}$ m^{-2}) and acidity in Scandinavia, and it is considered highly sensitive because of its granitic bedrock covered by thin and podzolic soils. An adjacent site, Birkenes II, only 6 km distant, has the same deposit, but is less sensitive because of the greater base weathering potential of its bedrock. Both can be compared with a pristine stream, Ingabekken, in central Norway, also considered sensitive but receives only about 0.5 g SO$_4$$^{2-}$ m^{-2}.

Bulk rain sulphate concentration at Birkenes is 63 μeq l^{-1}, of which about 89% is non-marine, while at Ingabekken it is only 19 μeq l^{-1} (and 42% non-marine). Stream sulphate concentrations (volume-weighted) are 131, 85 and 29 μeq l^{-1}

respectively, reflecting both the different site deposition and the degree of soil intervention. Acidity is highest in Birkenes I, 28 μeq l^{-1} (pH 4.55), while Birkenes II and Ingabekken scarcely differ (pH 5.01 and 5.10). High flow conditions at Ingabekken are characterized by low concentrations of calcium and high labile aluminium, and acidity is seen to be representative of pristine conditions during storm flow in southern Norway; equally Ingabekken could become acidified with additional acid deposition. Transient acidification is also reported for pristine sites in northern Sweden (e.g. Svartberget: Grip and Bishop, 1990; Bjarnborg, 1983) and in northeast North America (Murdoch, 1988).

North America: Sudbury lakes

Continued study of polluted lakes close to the Sudbury smelter in Ontario, Canada, shows various lakes' response to a halving of sulphur emission there during 1973–89 (Dillon *et al.*, 1987; Wright and Hauhs, 1991). At Clearwater Lake and others close to the smelter, sulphate deposition fell by 30%, lake acidity was greatly reduced and alkalinity recovered, even though base cations fell by about 25%; the fall in sulphate approximately matched that of emissions (Dillon *et al.*, 1986). The rapid recovery of this lake is attributed to the high weathering rate of the local soils/substrate which can provide sufficient base cations (Skeffington and Brown, 1993). The very much higher ($\times$ 100) sulphate and weathered base cation concentrations of the Sudbury lakes make comparison with lakes elsewhere acidified by distant sources rather unrealistic. A better comparison is provided by Plastic Lake and Harp Lake, about 100 km downwind in the Muskoka area, both classed as acidified in 1980–81, with pH values of 5.80 and 6.27 respectively; alkalinity in the lakes was low, 15 and 61 μeq l^{-1}. For these lakes there has been little response to emission reduction at Sudbury. Sulphate deposition had fallen by about 20% in 1989, but lake sulphate concentrations have not changed and lake alkalinity has decreased, implying their continued acidification (Wright and Hauhs, 1991). This lack of a positive response might be attributed to lack of close source/target links and to possible acid sources within the catchment, as well as to a lower weathering potential in the Muskoka area.

Hubbard Brook

In the northeast USA, studies in the forested Hubbard Brook experimental area also provide a long-term chemical data base (Driscoll *et al.*, 1989). Both sulphate and base cations in deposition have declined (by 30% and 50% respectively) over the period from the mid-1960s to the present. Sulphate and base cations in streamwater have also declined (by 20% and 12%) in parallel but neither pH nor alkalinity has changed.

Lake 223 and other experimental lakes (Schindler *et al.*, 1991)

Whole-lake experiments in North America have provided information about the process of acidification even though the timescale of change is inevitably more rapid, and their acidity has not reached them via their catchments. Lake 223 and Lake 302S (Ontario) and Little Rock Lake (LRL) (Wisconsin) have been progressively acidified by the direct addition of sulphuric acid from their natural circumneutral condition to pH values of 4.7 to 5.1. Lake 223 was sampled for two years prior to acid addition which was carried out over a six-year period (Schindler *et al.*, 1985). After maintaining acid conditions by further acid addition over three years, a controlled and phased recovery was then allowed (Schindler, 1987). Lake 302 has two almost identical basins, of which one (Lake 302S) received sulphuric acid annually over a decade, the other (Lake 302N) receiving an equivalent nitric acid addition. The latter basin was switched to hydrochloric acid after five years to assess non-biological sources of alkalinity (Kelly *et al.*, 1990; Rudd *et al.*, 1990). At Little Rock Lake, the pH was seasonally lowered during ice cover due to CO_2 from decomposition, but acid addition, beginning in 1985, brought the pH down to 5.2 for a two-year period after which a further fall to pH 4.7 was achieved. Pretreatment water chemistry in the three lakes indicated alkalinity at reasonable levels (25–80 μeq l^{-1}), calcium 44–110 μeq l^{-1}, and sulphate 53–74 μeq l^{-1}.

At each site, the estimated acid dose proved inadequate to reach the desired acid pH level, since alkalinity was generated internally in response to addition of strong acid. The source of alkalinity was microbial sulphate reduction and denitrification, as well as some algal uptake of nitrate; thus sulphur and nitrogen were retained within the lakes. Some release of calcium, magnesium and potassium from lake sediments provided additional bases. Overall, acidification disrupted the normal nitrogen cycle—nitrification was inhibited in the two Ontario lakes, while nitrogen fixation was reduced in Little Rock Lake. The aquatic phosphorus cycle was unaffected even though laboratory studies suggested that some changes might occur. The biogeochemical changes were consistent with those observed in lakes acidified by deposition. The rapid recovery phase of Lake 223, also attributed to internal alkalinity generation, suggests that acidification is reversible in these conditions. Biological changes during acidification and recovery of Lake 223 are discussed in more detail in later chapters.

Mount St Helens

Lakes in the blast zone of the volcano of Mount St Helens, Washington, which erupted in 1980, also provide an opportunity to study biogeochemical processes leading to recovery of such impacted systems. Some lakes were newly formed by the eruption, others changed in physical and chemical characteristics (Wissmar *et al.*, 1982a). Many of the changes were caused by the deposition of new mineral material in the blast cloud, some from erosional inputs. Only two lakes were slightly acidic (pH 6.17 and 6.21) after the event—although sulphate and chloride in one lake (Spirit Lake) increased by 157 and 85 times respectively, alkalinity was high (increased 22-fold) and many ions were enriched by the eruption. Lakes outside the blast zone had circumneutral pH and average alkalinity of 243 μeq l^{-1}, considered similar to the pre-eruption chemistry of the affected lakes. In addition, much of the buffering capacity was due to CO_2 evolved from the biotic degradation of dissolved organic carbon, associated with dense bacterial populations that developed (Wissmar *et al.*, 1982b).

7.4   TEMPORAL TRENDS

Chemical records

Historical data for lake chemistry have been collated to demonstrate progressive acidification in many areas. However, few provide consistency of methods of sampling and analysis, and these data also suffer from site relocation, gaps in the record, and other confounding changes. In many data sets, the record is too short to establish clear trends in pH that can be distinguished from flow-related, seasonal and annual variability. Even so, there is a strong conviction that many lakes have experienced a decline in pH over the past 100 years. The magnitude of the reported change varies from less than 1 μeq l^{-1} (less than the expected error of measurement) to as much as 100 μeq l^{-1} (i.e. 2 pH units from pH > 6.0 to pH ~ 4.0).

In Sweden (Sanden *et al.*, 1987), the best-documented and most reliable examples suggest a change from pH > 6.0 to < 5.0 from the 1940s to the 1970s, with little change since; following acidification fish populations declined and some species were lost (see Chapter 9). A 20-year record for Swedish rivers provides almost no evidence that their water quality has changed since the 1960s, even though Oden and Ahl (1972) found a mean pH decline of 0.25 pH units from 1965 to 1969 in 15 rivers (about 10% increase in acidity per year). (The addendum to their paper shows an unexplained 'spontaneous' increase in 1970, by as much as 0.6 pH units.) A 40-year record of a lake near Stockholm (Grimvall *et al.*, 1986) shows rising sulphate and calcium concentrations and a decline in alkalinity, but no significant change in pH. Although this period was one of S emissions rising to 1970, then decreasing, changes in sulphate concentration are not significant in relation to pH (Ahl, 1986; Sanden *et al.*, 1987); changes in runoff, reflecting rainfall and snow patterns, seem more important. In Lake Tjurken in central Sweden, sulphate doubled over 50 years, from 96 μeq l^{-1} to 181 μeq l^{-1}, while in southern Sweden it increased almost threefold in two lakes in the decade 1967–77, then declined (Ahl, 1986). During the phase of increasing sulphate concentration, the pH in one lake fell from 6.8 to 4.5 (H$^+$ increase > 30 μeq l^{-1}), but during the recovery phase when sulphate declined from 300 μeq l^{-1} to < 200 μeq l^{-1} a pH increase of only 0.2 units was seen (i.e. 10 μeq l^{-1}). Thus, although a strong coupling of pH and sulphate was seen during acidification, there is hysteresis in the deacidification phase; similar data are reported for other lakes in the region (Forsberg and Morling, 1987).

In Norway, acidification has been severe in the southernmost region, where fish stocks in an area of 33 000 km^2 have been damaged (Muniz, 1984). Comparing historical data from the 1920s and 1970s, an overall increase of 2.6 μeq H$^+$ l^{-1} occurred (Brown and Sadler, 1981), but 34 of 59 lakes showed only slight acidification, and 25 showed an *increase* of 7 μeq l^{-1}, of which 19 had a pH < 5.5 even at the time of the earlier sampling. The most recent '1000 Lakes Survey' (Henriksen *et al.*, 1987) was designed to examine changes since the 1974–75 survey. The lakes were in areas of 'sensitive' bedrock and restricted to those of < 20 ha so do not represent the population of lakes in Norway overall; clearly there is variability between individual lakes within the area. About 300 lakes were sampled in both the 1970s and 1980s surveys, and 100 lakes have been sampled each autumn since 1986 on a yearly basis, showing little change in their pH, although the number of lakes barren of fish has doubled and the area affected has increased by about 10% (Henriksen *et al.*, 1987, 1988). Sulphate concentrations were about 30% lower than in the 1970s in this region, while nitrate concentrations doubled; other chemical changes were a decrease in calcium and a rise in aluminium.

In Finland, records from 1965 to 1982 (Laaksonen and Malin, 1984) show a decline in water quality in 73% of sampled lakes, most during 1965 to 1970. A decline in pH was seen in the southern part of the Vuoksi river system, while a rising pH trend was seen in the northern Vuoksi basin. A north to south gradient of sulphate deposition is seen in Finland, matched by lake water sulphate, from 35 μeq l^{-1} in the north to > 100 μeq l^{-1} in the south.

In the UK, a searching analysis of 100 data sets for monitored river sites in sensitive areas was assessed (Ellis and Hunt, 1986); 75 had been analysed by the same methods and 50 included coincident alkalinity data. Acidification was not shown for most although six had a fall in pH (Fig. 7.5); none had a current pH < 5.0, and four were circumneutral. Some of the records showed a progressively increasing pH. In larger lakes, e.g. Windermere, alkalinity and pH have increased, possibly due to increased sewage input since nitrate and phosphate are higher in winter months. In the oligotrophic upland lakes and tarns in the Lake District, however, little chemical change is evident over the past 50 years (Sutcliffe, 1983), although diatom evidence suggests that four tarns (three very small) have become more acid (Battarbee *et al.*, 1988). Five streams in Wales draining both moorland and forested catchments have shown a decline in pH over two decades; only those draining forest fell to pH < 6.0 (Ormerod and Gee, 1990).

In the Adirondacks in the northeastern USA, lakes have been studied for some decades; 217 were sampled in 1929–37 and then again in the 1970s (Schofield, 1976). At the later date the lakes were judged more sensitive and some lacked fish. A survey in 1985 showed that, of more than 400 lakes, selected on a random basis, about a third had a pH < 5.0, and more than half an alkalinity < 40 μeq l^{-1}; about a third lacked fish (NYDEC, 1985). The recent NAPAP survey has incorporated this survey, finding 330 'sensitive' lakes of which 180 were considered acidified by acid rain (J. P. Baker *et al.*, 1990). Lakes of > 4 ha with low pH and ANC show a fall in ANC or non-marine Ca + Mg relative to sulphate and nitrate. Higher-ANC lakes increased in both pH and ANC, possibly due to catchment changes (Sullivan, 1990). However, a temporal trend in lake water chemistry cannot be established since methods of pH and alkalinity measurements changed (Pfeiffer and Festa, 1980). In Maine, six rivers have shown no significant change since 1935 (two rivers) or 1950 (four) (Haines and Baker, 1986; Stewart *et al.*, 1988) but some other areas in the northeast have changes consistent with some progressive acidification. Sensitive lakes in the upper midwest indicate increasing sensitivity, with high SO_4^{2-} relative to base cations. However, most lakes and streams in the USA have not experienced recent historical declines in pH or ANC. More quantitative estimates of acidification have now been made using palaeolimnological techniques (see below).

In Canada, lakes in Ontario draining the Shield geology are considered 'acidified' (ANC < 200 μeq l^{-1}) although their pH is rarely < 6.0. A survey of more than 200 lakes in the Sudbury region in 1974–76, repeated in 1981–83, showed that sulphate concentrations in lake water were related to the source of S emissions at the Sudbury smelter. Reduction of emissions by 1972, with improved dispersion via a 380 m high chimney, reduced deposition by about 50% by 1979, resulting in a 20–60% reduction in lake sulphate concentrations by 1989 (Dillon *et al.*, 1986; Wright and Hauhs, 1991). In Nova Scotia, data for 37 lakes and ponds draining base-poor geology showed a positive but non-significant geographical relationship between acidity and sulphate concentrations, and no change over time. For four Atlantic salmon rivers in the southern upland region, a significant fall in pH (about 0.5 units) was found over the 28 years from 1954–55 to 1980–81 (Watt *et al.*, 1983; Watt, 1987); this was associated with decline in alkalinity and rise in aluminium. For the Medway river, the pH decline was 0.02 pH units yr^{-1}. Nova Scotian waters are high in DOC, and lake acidity is affected by the drainage of organic acids from peatland catchments (Gorham *et al.*, 1986); their high humic content renders much of the aluminium non-toxic. Nonetheless, the three rivers of pH < 4.7 have lost salmon stocks, and 10 rivers of pH 4.7–5.0 have a reduced catch.

Diatom evidence of acidification

Reconstruction of past acidity of small lakes has been based on analysis of diatom and other biological remnants in lake sediments; it provides more convincing evidence of acidification over the past 200 years in Scandinavia, Britain and North America (Charles and Smol, 1991).

The relationship of the diatom assemblage to lake water quality takes account of acidity, alkalinity, calcium, aluminium and DOC (Birks *et al.*, 1990). The chemical components co-vary and are not independent of one another, leading some to doubt the validity of the inferred pH, since the diatom community may reflect both physical and other chemical conditions. Nevertheless, the relationships have been quantified and subjected to searching statistical analysis for a variety of waters, including the establishment of a reference data set for temperate diatom species.

The early method for diatom analysis was developed by Hustedt, based on tropical lakes, allocating diatom species to five well-defined groups:

— acidobiontic (Acb), at pH < 7.0, maximal at pH < 5.5
— acidophilic (Ac), at pH ⩽ 7.0
— indifferent (In), found equally at pH values above/below 7.00
— alkaliphilic (Alk), at pH ⩾ 7.0
— alkalibiontic (Alkb), at pH > 7.0.

Nygaard (1956) applied the Hustedt categories to lakes in Denmark, formulating an empirical index, α, to represent the composite diatom assemblage:

$$\alpha = 5(\%\text{Acb}) + (\text{Ac})/(\text{Alk}) + 5(\%\text{Alkb})$$

The arbitrary ($\times$ 5) weighting emphasizes the presence of Acb and Alkb groups, but in low pH lakes Alkb species are often absent, so the denominator may well be zero. Indifferent species were ignored.

Renberg and Hellberg (1982) developed another index, B, in which indifferent forms were included:

$$B = (\%In) + 5(\%Ac) + 40(\%Acb)/(\%In) + 3.5(\%Alk) + 108(\%Alkb)$$

Here even larger 'weighting factors' are involved! Index B is used to estimate a value for pH, based on a regression:

$$pH = 6.40 - 0.85 \log B$$

giving an accuracy of ± 0.25 pH units, similar to the accuracy of field measurement of pH (see Chapter 3).

These numerical and linear regression procedures have now been replaced by more rigorous procedures, using maximum likelihood, weighted averaging regression and regional calibration (Birks *et al.*, 1990) with an overall standard error of about 0.32 pH units for a set of 167 lakes. This technique overcomes some of the problems of earlier regressions and the rather subjective Hustedt categories.

Inferred pH history is now available for more than 60 lakes in northern Europe and North America. A consistent pattern of acidification over a 150-year period was claimed using Index B regression by Renberg and Hellberg (1982). A closer examination of the record shows substantial differences in the timescales of acidification of adjacent lakes, as indeed are revealed by current chemistry. Influences other than pH on diatom assemblages and other communities are known, such as lake trophic state, silicate thresholds, toxic metals and organic materials, as well as physical conditions such as temperature (Psenner and Schmidt, 1992). The statistical procedure of canonical correspondence analysis has shown potential for following up some of these (Charles and Smol, 1991). The precise dating of cores by markers such as lead or caesium isotopes, and analysis of the sedimentation record, are required to establish precise timescales of change. Although such techniques are well known, they are not always applied, and single dates (e.g. from the ^{137}Cs fallout peak) are sometimes extrapolated regardless of changes in sedimentation. Further weaknesses are that historical pH inferences can seldom be validated by independent data since chemical records are few and unreliable, are usually for larger lakes than those sampled for palaeolimnology, and are often subjected to other unrecorded changes. The potential of palaeolimnology was reviewed objectively by Davis and Smol (1986); some but not all of the problems they identified have since been overcome. A substantial fraction of lakes studied (Charles *et al.*, 1989) were historically acidic (pH < 5.5), attributed to the geological or organic catchment conditions of some regions; in some studies only the more acidic lakes, some due to natural conditions, were selected. Analysis of a wider variety of water bodies is desirable.

A stimulus to the long tradition of palaeoecology was provided from interest in acidification and its causes, especially since historical chemical records are insufficient. The initiation of the 'Surface Water Acidification Programme' (SWAP) provided a further stimulus (and funding) for coordinated studies between the UK and Scandinavian countries (Battarbee *et al.*, 1990; Mason, 1990); parallel studies were undertaken in North America. Related investigations at several sites provided information regarding current deposition, soil and water chemistry, and aquatic biology. Several advances in technology were achieved, allowing retrospective analysis of environmen-

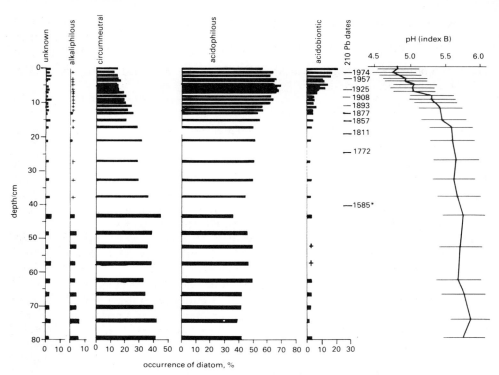

Fig. 7.8—Changes in diatom groups in sediment cores and inferred pH for a lake in Galloway, southwest Scotland (after Battarbee *et al.*, 1989).

tal changes and critical evaluation of alternative hypotheses of causes of acidification. For a few sites, recent changes have been related to a longer-term retrospective over post-glacial time (Renberg and Battarbee, 1990).

Studies of post-glacial sediments show that acidification occurred in the past 10 000 years at a slow rate—up to 0.1 pH unit/1000 years—until about 1850. Acute acidification occurred, in contrast, some time after 1800 in sensitive sites receiving high levels of acid deposition (Fig. 7.8). Further, it is found that recent sediments of acidified lakes are contaminated also by trace metals and carbonaceous particles, some associated with fossil fuel combustion. Analyses of pollen, chrysophyte scales, algal pigments, crustacean and insect remnants have also been used in studies to reconstruct past conditions, not only of acidity but also of salinity, eutrophication and even community interactions. The potential for hindcasting to past conditions is now substantial, and to a degree has been validated by modelling (see Chapter 11).

In western Britain (Battarbee *et al.*, 1989) a number of acid upland lakes has been sampled. Most have current pH < 5.5 and calcium < 50 μeq l^{-1}; they lie in areas of high deposition and several have poor or absent fish populations. Major changes in the diatom flora have occurred since about 1850 when pH inferred from the diatom record was about 6.0, but two had pH < 5.5 at that time. The largest pH change is for Welsh lakes, as large as 1.5 pH units; most lakes studied there were acid even 100 years ago but a sharp increase came around 1940. In contrast, in Scotland most studied lakes did not become acid until the turn of the century. Indeed, some seem to have changed only recently even though they are highly susceptible by virtue of their high rainfall, hard geology and organic and base-poor soils.

In the NAPAP study, palaeolimnological analysis has been applied to a selection of the sampled lakes (Sullivan, 1990) to provide quantitative estimates of the rate of acidification. For Adirondack lakes the loss of ANC for currently acidic lakes has been 18 μeq l^{-1} since 1850, with an equivalent increase in aluminium; for the most sensitive, lowest ANC lakes these changes have been less marked, with a loss of 7 μeq l^{-1} ANC and an increase of 1 μeq l^{-1} for aluminium.

7.5 TIMING OF ACIDIFICATION

Overall, the time of recent acidification inferred from sediment cores shows a range from about 150 years ago to quite recent (1975) changes (Table 7.2). In Scotland, Galloway lakes began acidifying in about 1840, but some in the region are not acidified and others have become so only recently (e.g. Loch Fleet) (Anderson *et al.*, 1986). Cumbrian lakes and tarns are thought to have become acid 5000 to 10 000 years ago, and have shown little recent change (Sutcliffe, 1983). Welsh lakes became acid rather consistently about 1850 (Battarbee *et al.*, 1988). Three lakes in southern Norway changed by 0.5 to 0.8 units in 1890, 1918 and 1927, while in Sweden and Finland acidification was more recent, after 1945 and 1960 respectively. At Gardsjon, however, slower changes are evident over a much longer timescale, 12 500 years. In Germany pH decreases are reported in the past 30 to 40 years and a similar timing applies to the Netherlands, Belgium and Denmark. In North America, Canadian lakes have become acid over the last 70 years, whereas lakes in the northeastern USA changed only recently, mostly since about 1959 (Harter, 1988). A consistent relationship is not seen between present pH and the onset of acidification, which might be expected from Henriksen's 'titration' concept of acidification (see Chapter 6), with exhaustion of buffer materials from within the catchment. Differences between in-lake alkalinity generation, as well as altitude and regional temperature regimes, have been invoked to explain the different timings for two Alpine lakes (Psenner and Schmidt, 1992). Recent or current acid loading is not matched by the onset or degree of acidification. Catchment conditions (land use, hydrology, vegetation) and local acidifying sources still seem important in the differential response of water bodies to regional patterns of deposition.

7.6 UNRESOLVED ISSUES

Selection of small acidic lakes to represent a regional distribution is flawed; few surveys have taken steps to provide a statistical estimate of their regional distribution. Similarly the historical perspective from palaeoecology has been based on such lakes.

The degree of acidity causing concern is defined differently in different countries, making generalized comment difficult; consistent definitions would be helpful.

Table 7.2 Diatom evidence for the timing of recent acidification for some European and North American lakes

Lake, location	Current pH	Date acidified	Pre-acid pH	Inferred pH	Δ pH
Norway:					
Brärvatn	5.2	1850	5.3–6.1	5.2–5.3	0.3
Hovvatn	4.4	1918	4.8–5.4	3.9–4.4	0.75
Holmvatn	4.7	1927	4.8–5.2	4.9	0.5
Blävatn	5.1	1930	5.2	5.1	n.s.
Verevatn	4.4	1940	5.0	4.5	0.5
Sweden:					
L. Oresjön	4.6	1900	6.1	4.6–4.7	1.5
Gärdsjön	4.6	1950	6.0	4.5	1.5
Lysvatten	5.9	—	6.1	5.2	0.9
Finland:					
Hauklampi	4.8	1961	6.0–6.4	5.1–5.4	1.05
Germany:					
H. Wiesersee (Black Forest)	3.6–4.2	1954	4.6–4.7	3.8	0.9
United Kingdom:					
Enoch	4.5	1840	5.3	4.4	0.9
Llyn Lagi	4.9	1850	6.1	4.6	1.5
Scoat Tarn	5.0	1850	6.0	4.6	1.4
Dee	5.6	1890	6.1	5.6	0.5
Devoke Water	6.3	1900	6.6	6.1	0.5
Grannoch	4.7	1930	5.7	4.7	1.0
Skerrow	5.9	—	5.9	5.8	n.s.
North America:					
Big Moose, NY	4.7	1950	5.1	4.7	1.0
Woods, NY	4.7	—	5.0–5.5	4.7–4.9	0.5
Honnedaga	4.7	1966	6.2	5.6	0.6
Clearwater	4.3	1930	6.0	4.2	1.8
CS Lake, Ont.	5.2	1954	7.1–7.3	6.4	0.8
B Lake, Ont.	5.2	1962	6.2	4.7	1.5

'Current pH' is that recently measured; 'pre-acid pH' and 'inferred pH' are estimates from diatom analysis of cores; Δ pH is the change inferred by the diatom analysis; 'n.s.' is non-significant.

Alkalinity is a more meaningful measure of acidification than pH, but has not always been measured.

The variety of response, even between lakes on similar terrain and receiving the same deposition, suggests that local factors, including other sources of acidity, play a role.

In particular, soil and hydrological conditions in catchments, and the rate of weathering, is not sufficiently documented.

Present and historical levels of acid loading are often not known on less than a regional scale, making it difficult to match the rate of changes in surface waters with changes in emissions.

In-lake processes are also evidently important, but are not yet well quantified or developed as a general concept.

Historical chemical records are few and limited in their scope; they can seldom be matched with current measurements and provide only a poor indication of trends.

Studies of fossil remnants in sediments are helpful, but a greater variety of lakes and their associated physical and chemical conditions would broaden the base for generalization.

7.7   SUMMARY AND CONCLUSIONS

Several regions in Scandinavia, Europe and North America have a proportion of surface waters which are acidic, i.e. pH < 5.5.

There is undoubted evidence that surface waters in sensitive areas have become acidified over time, now more clearly demonstrated by improved palaeolimnological techniques.

The degree of acidification over time varies from $1 \ \mu eq \ l^{-1}$ to $100 \ \mu eq \ l^{-1}$, and time for acidification from 5 to 150 years. In only a few cases is the change in acidity matched with changes in sulphate concentrations, and such changes do not necessarily correspond to the onset or timing of increased atmospheric loading.

Acidification processes involve soil and bedrock conditions in the catchment, as well as rainfall and runoff routes and rates; the relationship of the yield of buffering materials in response to atmospheric input is only now being quantified.

In some catchments with a reserve of ANC, short-term acid episodes may occur during heavy rain, or with sea salt rains, or after snowmelt when soil contact is minimal. These conditions are seen even in areas with little atmospheric acid burden.

Historical records of chemical acidification are almost never available for a sufficient period or with sufficient consistency to document a trend in acidification. Fossil diatom and other evidence from sediment cores from about 70 small lakes provide evidence for acidification over the past 150 years, although this does not always match recent (or historical) acid loading.

8

The role of aluminium and other metals

8.1 INTRODUCTION

Aluminium is ubiquitous in soils and surface waters. It occurs in higher concentrations in acid waters, and is often invoked to explain fish absence or fish kills at otherwise tolerable levels of acidity. Other transition metals, such as iron and manganese, and copper, cadmium, lead, chromium and zinc, are also more soluble or possibly more toxic in acid waters. They are more prevalent in areas with mineral-containing ores. The role of these trace metals in limiting or restricting species of the aquatic community is uncertain, although it is clear that, with the exception of manganese, they can be toxic if in sufficient concentration (Andersson and Nyberg, 1984; McDonald *et al.*, 1989; Turnpenny *et al.*, 1987; J. P. Baker *et al.*, 1990). All these metals, including aluminium, affect ion regulation and respiratory function of the fish gill. Aluminium, cadmium and manganese also affect calcium metabolism, development and skeletal calcification, as well as having long-term effects on spawning and recruitment. It is recognized that potentially toxic metals are made more mobile in acid soil conditions, but in acid surface waters their toxicity, for example for copper, cadmium and zinc, is not always greater than in circumneutral waters, perhaps due to the competition of H^+ for binding sites on the gills (McDonald *et al.*, 1989).

Aluminium is the metal most investigated in acid systems. It is released from acid soils in which the base reserve has been exhausted, first replacing calcium and magnesium on the soil exchanger (see Chapter 5), along with hydrogen ions, then being mobilized to soil drainage by the excess anions present. This leads to sustained or episodic release to fresh waters where aluminium is toxic to fish and other organisms within the limited range of pH characteristic of most oligotrophic waters. The relative toxicity of aluminium and other trace metals is related to their chemical speciation, principally determined by pH.

It may seem strange that aluminium should have such deleterious effects—it is the third most abundant element in the Earth's crust, and the world's most common metal, it provides vessels for food preparation because it is inert, and it is used for water purification and for common pharmaceuticals. The explanation lies in its chemical transformations.

The free metal is not found in nature because it is highly reactive, but aluminium forms are found in all geological materials and it is transferred in some degree to all living organisms. However, it is not an essential element and has no known function in biological systems. As a result of prolonged weathering of the Earth's primary minerals, especially feldspars and micas, aluminium is found predominantly in sedimentary clays such as montmorillonite and kaolinite. When silica is leached from these materials it leaves behind aluminium hydroxides such as gibbsite ($Al_2O_3.3H_2O$) or boehmite. The principal aluminium ore is bauxite, containing hydrated aluminium compounds and also some iron oxide. Weathering may lead to the formation of sulphate minerals such as alum or aluminate.

8.2 ALUMINIUM CHEMISTRY

Aluminium in solution is amphoteric, reacting with both acids and alkalis; it is extremely reactive and the free metal is not found in nature. These properties encourage the formation of a variety of chemical species and complexes, depending on ambient pH, temperature, and other chemical constituents. Aluminium always exists in the trivalent state, and although metallic, it exhibits marked covalent tendencies, forming very stable complexes with a variety of inorganic and organic materials. Inorganic compounds include a number of identifiable chemical 'species'; in fresh waters of pH 5 to 9, these species are distinct chemical entities, several being found together within this pH range (Fig. 8.1). The trivalent Al^{3+} occurs only at pH < 6.0, where two hydroxides, $Al(OH)^{2+}$ and $Al(OH)_2^+$, are also present. Above pH 6, $Al(OH)_4^-$ is increasingly evident as pH rises. A fourth hydroxide, $Al(OH)_3$, is maximal at around pH 6. At a characteristic range of pH values of 5 to 6 in oligotrophic waters four species coexist and are included in the operational fraction as 'inorganic monomers'. Below pH 5, the trivalent Al^{3+} increases progressively, and at pH 4 it is the major, or even the only aluminium ion present (Stumm and Morgan, 1981). This scheme is based on the assumption that the solution in fresh water is in equilibrium with the mineral gibbsite, a questionable view since other aluminium minerals may prevail. Secondly, $Al(OH)_3$ has not been identified specifically in natural soil and surface waters, even though it is included in most models (Farmer, 1986; Glover, 1987).

In attempting to define the toxicity of aluminium in natural waters it is important to know what forms are present and their toxicity, as well as stable, non-toxic, complexes with various ligands which might moderate toxicity. Complexing agents include inorganic ions such as fluoride, sulphate and silicate (Birchall et al., 1989), and organic materials such as humic and fulvic acids (Petersen et al., 1987) and citrate (Baker and Schofield, 1982). Even where the possible species and rate constants are known, it is unlikely that equilibrium will prevail, especially in the variable, short-term, conditions of low-order streams.

In practice, aluminium in water samples is defined by operational procedures; the scheme devised by Driscoll (1984) is that most often followed. '**Total aluminium**' can

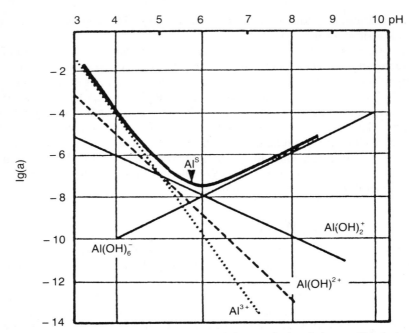

Fig. 8.1—The activity of soluble inorganic aluminium species in relation to pH (calculated from thermodynamic data at 25°C) (after Seip *et al.*, 1984). Al^s is the sum of the activities of aluminium species in solution.

be found from an unfiltered sample after acid or ultraviolet digestion followed by analysis. A filtered ($< 0.45 \mu$m) sample contains both inorganic and organic complexes but excludes most polymeric, colloidal or solid particulates. It is extracted by 'oxine' or catechol violet and then measured as '**extractable or reactive aluminium**', or '**total monomeric aluminium**'. Separation of this into '**inorganic monomeric**' and '**organic monomeric**' fractions is achieved by extraction with a chelating resin (Fig. 8.2; Driscoll, 1984). The former includes Al^{3+}, AlF_3 and Al hydroxides; the latter includes organic complexes. At pH < 6.5 this is effective since most of the inorganic fraction, thought to include the toxic species, is dominated by positively charged ions.

Techniques differ between laboratories, so estimates of the fractions (and their associated aluminium species) tend to differ. Further, the widely variable ionic strength and level of organic materials in surface waters leads to some differences. While total aluminium perhaps represents a potential for toxic effects, it is the inorganic monomeric fraction which contains the toxic hydroxy species, but opinions differ as to which is the most potent, possibly also including polymeric species (see below). This fraction may be only a few percent of the total (depending on other water components).

In addition to the problems of matching the operational fractions with toxic species, it should be recognized that samples are not always stable over the time between collection and analysis. A change in temperature (Lydersen *et al.*, 1990), for

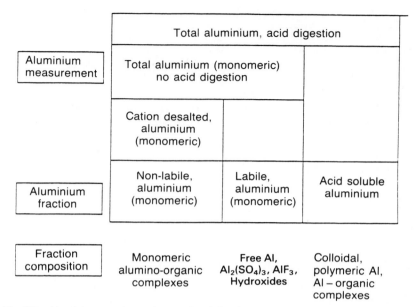

Total aluminium, acid digestion		

<table>
<tr><td>Aluminium measurement</td><td colspan="2">Total aluminium (monomeric)
no acid digestion</td><td></td></tr>
<tr><td></td><td colspan="2">Cation desalted,
aluminium
(monomeric)</td><td></td></tr>
<tr><td>Aluminium fraction</td><td>Non-labile,
aluminium
(monomeric)</td><td>Labile,
aluminium
(monomeric)</td><td>Acid soluble
aluminium</td></tr>
</table>

Fraction composition	Monomeric alumino-organic complexes	**Free Al, $Al_2(SO_4)_3$, AlF_3, Hydroxides**	Colloidal, polymeric Al, Al – organic complexes

Fig. 8.2—Aluminium species and operational fractions following the Driscoll separation method.

Table 8.1—Major chemical ionic species of trace metals as a function of pH (after McDonald *et al.*, 1989)

Metal	pH 7.5	pH 5.0	pH 4.0
Copper	$CuCO_3$, Cu^{2+}	Cu^{2+}	Cu^{2+}
Cadmium	Cd^{2+}, $CdOH^+$	Cd^{2+}	Cd^{2+}
Zinc	$ZnOH^+$, Zn^{2+}	Zn^{2+}	Zn^{2+}
Lead	$PbCO_3$, $PbOH^+$	Pb^{2+}	Pb^{2+}
Aluminium	$Al(OH)_4^-$	$AlOH^{2+}$, $Al(OH)_2^+$	Al^{3+}
Manganese	Mn^{2+}	Mn^{2+}	Mn^{2+}
Nickel	Ni^{2+}	Ni^{2+}	Ni^{2+}
Iron	$Fe(OH)_2^+$, $Fe(OH)_4^-$	$Fe(OH)_2^+$	$Fe(OH)^{2+}$

instance, can shift the distribution of chemical species, and an increase in temperature can induce over-saturation and precipitation (Seip *et al.*, 1984). Similarly, CO_2 degassing may raise pH and have the same effect (Tipping *et al.*, 1988). A practical compromise is to carry out preliminary sample procedures in the field, followed by rapid analysis.

Other metals found in acid waters do not have the same chemical complexity, although for some the ionic species differ with pH (Table 8.1). Standard methods of analysis are sufficient for assessment of their toxicity in acid systems.

Table 8.2—Chemical reactions within soils (after Tipping, 1989)

Chemical reaction	Pathways, processes
$NH_4^+ \rightleftharpoons NH_3 + H^+$	NH_3 taken up by plants
$NO_3^- + H^+ \rightleftharpoons HNO_3$	HNO_3 taken up by plants
$CO_2 + H_2O \rightleftharpoons H_2CO_3$	In deeper horizons
$H_2CO_3 \rightleftharpoons H^+ + HCO_3^-$	In stream waters
$Al(OH)_3 + 3H^+ \rightleftharpoons Al^{3+} + 3H_2O$	Dissolution/precipitation of $Al(OH)_3$;
$Al^{3+} + H_2O \rightleftharpoons AlOH^{2+} + H^+$	hydrolysis reactions of Al in deeper
$AlOH^{2+} + H_2O \rightleftharpoons Al(OH)^{2+} + H^+$	horizons, in stream water
$Al(OH)_2^+ + 2H_2O \rightleftharpoons Al(OH)_4^- + 2H^+$	

$$Na \rightarrow Na^+$$
$$K \rightarrow K^+$$
rock
$$Mg \rightarrow Mg^+$$
$$Ca \rightarrow Ca^+$$

Weathering in base rock and mineral horizon

8.3   ALUMINIUM IN SOILS AND SEDIMENTS

Metastable weathering products are the principal sources in soils of mobilizable aluminium and silicon—they are common in the acid soils of temperate and boreal climate zones, and elsewhere in cool mountainous regions. The reprecipitation of aluminium released by weathering is part of the process of podzolization which leads to separation of the organic and mineral soil horizons, influencing both the pathways of soil drainage and the replenishment of base cations in the leached organic layer of soils. Aluminium in the upper soil horizons is mobilized by acidity by naturally generated organic and inorganic acids together with that deposited in rain, but precipitates in the less acid lower mineral horizons (Bache, 1984; Farmer, 1986). With low pH (< 5.0) in the upper soil, H^+ ions in the soil water are absorbed to aluminium hydroxides on the surface of soil particles (the 'soil exchanger') and are exchanged for Al^{3+} ions. When the Al^{3+} ions are released into the soil water by acidic anions (sulphate, nitrate, chloride), they capture OH^- ions to form free hydroxy-Al ions and reduce the pH, but some are reprecipitated as insoluble basic aluminium sulphate, $Al(OH)SO_4$. Some relevant processes in soils are summarized in Table 8.2. In the deeper, less acid horizons, aluminium tends to form colloidal hydroxy-Al species and to precipitate as aluminium humates and fulvates, as hydrated aluminium silicate (imogolite) or as polymeric hydroxyaluminium species. In the B horizon leachate the silicate species have a limited Al : Si ratio of about 0.3 so these complexes do not predominate (Farmer, 1986). Mobilized iron is also precipitated as humates and as a hydrated iron oxide, ferrihydrate.

Soils from different sites are clearly different in character and in their potential to leach aluminium—some, derived from schists and granites, are able to yield large

Table 8.3—Aluminium and silicon in soil leachates and runoff for four sites (after Farmer, 1986)

Horizon	Site 1 Al/Si	Site 2 Al/Si	Site 3 Al/Si	Site 4 Al/Si
Humus	1.0/3.1	0.19/1.4	0.3/0.6	0.89/0.67
E/A2	0.65/2.0	0.34/1.5	1.7/4.0	3.9/2.9
Lower B	0.65/2.2	0.27/3.1	0.6/3.1	0.98/0.71
Lake/stream	0.71/1.0	0.18/2.8		

Sites 1 and 2 are at Hubbard Brook (New Hampshire)—a hardwood forested catchment. The soil podzol is developed on a sandy loam, soil pH 4.4 in the B horizon, 4.2 in C. Rain pH is 4. Soil acidity is in part from oxidized mineral sulphides. Al is precipitated in lower B as Al-organic complex and most Al is leached from the humus layer.
Sites 3 and 4 are in Vosges, France, with a podzol (3) and a brown soil (4), both lying above sandstone. Rain pH is 5. Silicon is deposited in the B horizon but some Al is leached.

fluxes of aluminium and silicon due to weathering. In areas subjected to acid rain (pH 4.0) aluminium and silicon draining from the B horizons are little higher than those from soils at a site with rain of pH 5.0 (Farmer, 1986; Table 8.3). While all the soils in this table demonstrate aluminium leaching from the B horizon, the highest is from a forested site with a rain pH of 5.0 with an acid brown soil, not a podzol. In contrast, with tropical and temperate podzols on coastal sands, aluminium humates are deposited at the boundary of upper with lower horizons. While these deposits can persist for thousands of years in circumneutral conditions, they can be mobilized by acid leaching, as seen in Florida sites, with soil pH < 5.0 (Farmer, 1986).

Several mechanisms have been proposed for the solid-phase control of aluminium concentrations in the soil solution. Calculated values of aquo-aluminium in soil solutions are consistent with those predicted for basic aluminium sulphate (van Breemen, 1973, 1991), supporting the claim that atmospheric sulphate deposition has acidified the soils and transformed aluminium oxides to the basic sulphates (Eriksson, 1981; Reuss and Johnson, 1986). However, sulphate also arises from oxidation of sulphides in recent marine sediments, or more slowly from weathering of sulphide-rich rocks derived from such sediments. Soils in Wales, derived from mudstones (Farmer, pers. comm.) and those in the Adirondacks in the northeastern USA (Driscoll *et al.*, 1984) are acidic, probably for this reason. It is now doubted that aluminium sulphate minerals (jurbanite, alunite, basaluminite) control aquo-aluminium levels in soils, since these are not confirmed by X-ray analysis (Driscoll *et al.*, 1984; Driscoll and Schecher, 1988). Concentrations of H^+ ions are now considered to be sufficient to explain aluminium reactions in the soil solution (Glover, 1987) along with the excess of acid anions ('**mobile anions**' of Johnson and Cole, 1980). It could be argued, however, that long residence times in soils may allow some species such as alunite to be formed owing to the persistence of metastable conditions over several years.

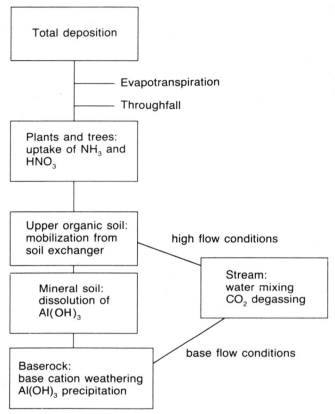

Fig. 8.3—Chemical pathways and processes involved in the release of aluminium to stream waters (after Tipping, 1989).

8.4   ALUMINIUM TRANSFER AND SPECIES IN SURFACE WATERS

The extent to which the aluminium leaches through the soil profile depends on the hydrological pathway—where there is lateral leaching above the podzol 'pan' aluminium humates may be released to surface water, whereas that draining through the lower B horizons carries labile inorganic aluminium species. The vegetation (e.g. moorland versus forest) and the extent of canopy closure of forest influences the amount of humic materials present in soils. Forest drainage is typically higher in aluminium than that from moorland (Krug and Frink, 1983; Miller, 1984; Ormerod et al., 1989; Reynolds et al., 1988). This is reasonably related to the additional input of acid anions through forest canopy scavenging and enhanced cation exchange within the soil (Adams et al., 1990). Major processes and pathways are summarized in Fig. 8.3.

Lake and stream sediments may contain significant reserves of aluminium and are evidently important in influencing surface water concentrations. Precipitation of aluminium to, and dissolution from, the sediment occurs as water quality changes, for example during changes in dissolved oxygen concentration or during acid episodes associated with snowmelt or heavy rain. This phenomenon may explain the rapid appearance of aluminium in streams following snowmelt when the enhanced acidity may mobilize aluminium from stream/lake sediments (Johnson, 1979; Kullberg and Petersen, 1987). Alternatively, if soil water is displaced by snowmelt or by an influx due to heavy rain, the aluminium may be mobilized within the soils even though conditions there are more stable. Another reason for fluctuating aluminium concentrations and species occurs when waters of different compositions meet (Rosseland *et al.*, 1992); 50% of salmon and sea trout in the mixing zone of a river were killed within 7 hours, attributed to rapid changes in aluminium speciation.

Several such episodes have been reported in the literature (e.g. Driscoll *et al.*, 1984; Hooper and Shoemaker, 1985; Sullivan *et al.*, 1986; Wigington *et al.*, 1990), confirming a strong relationship of surface water quality with soil water composition and drainage pathways during episodes. Seasonal changes also result in fluctuations in aluminium species (Fig. 8.4; Driscoll *et al.*, 1984); at a site in New Hampshire, low pH and high flows associated with summer rains led to high levels of inorganic monomeric aluminium, whereas low flows in winter led to raised pH and higher aluminium. TOC and organic aluminium were generally high in autumn and summer, low in winter and spring. Overall, high inorganic aluminium is associated with high acidity, high flows and high anion content.

Although aluminium is such a common element in soils, and mobilized in acid conditions, concentrations in fresh waters are usually very low, within a range of a few parts per million (mg l^{-1}, even < 100 μg l^{-1}); this is because of its low solubility and tendency to form less soluble complexes with natural materials present in soil and surface waters. In waters in the pH range 6–9, aluminium solubility is very low, but in waters of pH < 5.0, total aluminium may be much higher, especially in oligotrophic waters where complexing agents are limited. In natural waters the inorganic monomeric aluminium appears to exist as Al^{3+} and soluble complexes with OH^-, F^- and SO_4^{2-}, while slower-reacting species include oxyhydroxides and aluminium silicates. In a variety of Swedish waters, Lydersen *et al.* (1990) calculated that at pH 5.0 and 25°C, Al^{3+} is 36%, $Al(OH)^{2+}$ 37%, $Al(OH)_2^+$ 26% and $Al(OH)_3$ 1%; at the same pH at 2°C, these species are present at 84%, 13%, 2% and 0% respectively. Doubling acidity or reducing temperature yields the same change in $Al(OH)_3$. At pH 6, organic fractions predominate. Humic and fulvic acids are strong complexing agents in fresh water, reducing the proportion of toxic ionic species. This is clearly evident in acid Nova Scotian waters where as much as 90% is associated with organic components (Lacroix and Kan, 1986); the dissolved organic content of these waters is commonly in excess of 10 mg l^{-1}, and fisheries decline there is not attributed to aluminium toxicity.

There is, however, still uncertainty regarding the chemical species present. Analysis of the relationship of Al^{3+} and H^+ in the Norwegian '1000 lakes' survey (Fig. 8.5; Henriksen *et al.*, 1988a) showed that there is no simple relationship which might suggest equilibrium with $Al(OH)_3$ (Seip *et al.*, 1990). This might be expected from knowledge that waters draining laterally from upper horizons will have a high

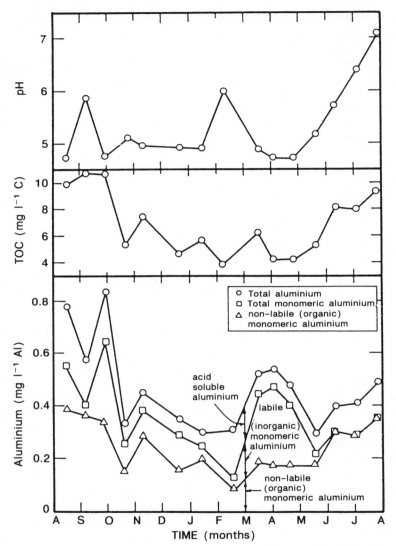

Fig. 8.4—Seasonal changes in aluminium species in an acid stream (after Driscoll *et al.*, 1980).

proportion of organic complexed aluminium, while drainage from lower B horizon soils will be higher in inorganic aluminium. Some enlightenment comes from field experiments where aluminium has been mobilized by addition of acid (H^+) to stream systems where some is neutralized by the release of aluminium from suspended particulates, the stream bed, submerged mosses or liverworts (Henriksen *et al.*, 1988b; Norton *et al.*, 1987); with an acid pulse of pH 4.5, about 0.5 g Al m^{-2} from the stream bed was released over 44 hours and stream concentrations of aluminium rose to

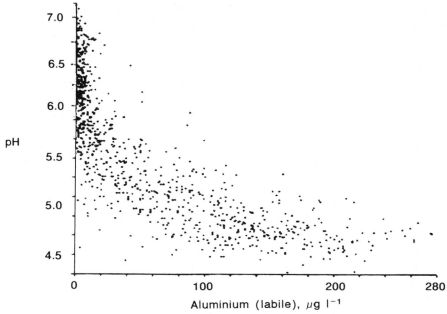

Fig. 8.5—Relationship of measured [Al] total and pH in 1000 lakes in Norway (after Henriksen *et al.*, 1988).

2.5 mg l^{-1}. It has been shown that the liverwort *Nardia compressa* and *Sphagnum* are able to retain aluminium and to release it in acid conditions (Caines *et al.*, 1985; Clymo, 1984; Weider *et al.*, 1988). In a laboratory experiment, Tipping and Hopwood (1988) found that a mineral stream bed subjected to pH 4.5 could release less than 28 μg m^{-2} s^{-1}, while for a liverwort-rich stream bed aluminium release was as much as 1.7 mg m^{-2} s^{-1}.

8.5 ALUMINIUM TOXICITY

Most studies of aluminium toxicity in fresh waters have been on fish, mostly sensitive salmonid species. At circumneutral pH and oligotrophic water, low concentrations (50–70 μg l^{-1}) disturb the sodium balance of blood and tissues by affecting the active uptake of sodium across the gills (McDonald *et al.*, 1989; Potts and McWilliams, 1989). The principal mechanism of response is the alteration in influx and efflux of ions (both Na$^+$ and Cl$^-$) across the gill membrane, with low blood sodium resulting either from impaired uptake at low pH (Dalziel *et al.*, 1986) or from enhanced efflux (Booth *et al.*, 1988). The toxicity of aluminium is independent of effects of pH *per se* and is most significant at pH 5.0–6.0 (Howells *et al.*, 1990). At higher aluminium concentrations and pH ($\geqslant$ 6.5) where Al(OH$_4$)$^-$ is present, respiration (Wood and McDonald, 1987) is affected by excessive mucus production, perhaps due to the discontinuity of conditions at the gill/water interface, reducing aluminium solubility, but anoxia is not directly responsible for mortality. The ultimate cause of death in trout exposed to low pH, low calcium and high aluminium conditions, can be attributed to a syndrome of physiological and morphological responses (Wood, 1989); effects at the gill membrane have been illustrated schematically in Fig. 8.6. A variety of ameliorating factors are known to influence aluminium toxicity—they include ambient concentrations of calcium (Brown and Lynam, 1981), silicate (Birchall *et al.*, 1989), fluoride, citrate (Baker and Schofield, 1982) and humic substances (Lacroix and Townsend, 1987). The degree to which these substances 'protect' against aluminium toxicity in the field is limited by their concentrations—often low in oligotrophic waters and in many cases not reported.

The sensitivity of fish is significantly different between species or even strains of the same species, as well as between different life stages. The developing embryo, young fry and migrating stages of salmonids are the most sensitive. Other freshwater species such as perch, pike and whitefish are relatively tolerant, but early life stages of roach are sensitive (Vuorinen *et al.*, 1992a), perhaps explaining loss of roach populations in lakes with pH ≥ 5.5 in central Sweden.

Sublethal effects of aluminium have also been reported; they include reduced growth (Sadler and Lynam, 1987, 1988), slow skeletal development (Reader *et al.*, 1988; Dalziel and Lynam, 1992), reproduction (Vuorinen *et al.*, 1992b), and anomalous behaviour (Cleveland *et al.*, 1986).

In general, the concentrations of trace metals (Pb, Cd, Cu, Zn, Mn, Fe) in acid waters are not thought to be sufficient alone to lead to toxic effects on fish (McDonald *et al.*, 1989; J. P. Baker *et al.*, 1990), although they might be expected to impair gill function. In the field, however, they may occur together, and in acid, low-calcium, waters their combined action seems to influence fishery status and the variety of species present (Turnpenny, 1987, 1989).

8.6   MODELLING ALUMINIUM TOXICITY FOR FISH

Modelling and predicting fish populations exposed to acid and aluminium have been attempted by the NAPAP study (J. P. Baker *et al.*, 1990) on the basis of laboratory bioassays. These have been documented for (a) sensitive, (b) intermediate and (c) tolerant North American species, specifically rainbow trout (*Oncorhynchus mykiss*), smallmouth bass (*Micropterus dolomieui*), and brook trout (*Salvelinus fontinalis*). Empirical relationships between pH, calcium and aluminium were found, as follows:

(a) 21-day rainbow trout swim-up fry:

$$x = -8.90 + 1.56 \, \text{pH} + (4.08 \times 10^{-3})\text{Ca} - (7.04 \times 10^{-3})\text{Al}$$

(b) 8-day smallmouth bass alevins:

$$x = -18.73 + 3.57 \, \text{pH} + (1.45 \times 10^{-3})\text{Ca} - (4.37 \times 10^{-3})\text{Al}$$

(c) 21-day brook trout swim-up fry:

$$x = -23.49 + 5.35 \, \text{pH} + (2.97 \times 10^{-3})\text{Ca} - (1.93 \times 10^{-3})\text{Al}$$

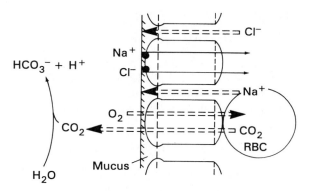

A. Structure and Function of Unstressed Fish Gill
- mucus coats external gill surface at water surface
- chloride cells 'pump' Na^+ and Cl^- from water into fish
- Na^+ and Cl^- leak from fish between cell junctions to water
- red blood cells (RBC) exchange CO_2 for O_2 inside gill
- O_2 and CO_2 diffuse across epithelial cells of gill
- CO_2 reacts with water to form HCO_3^- and H^+ outside of gill

B. Decreasing pH (Increasing Acidity)
- displaces Ca^{2+} from gill surface
- leads to increased gill damage
- increases rates of Na^+ and Cl^- loss
- inhibits active Na^+ and Cl^- 'pumping'
- causes ionoregulatory toxicity

C. Increasing Aluminium at Low pH
- aluminium binds to gill surface
- further displaces Ca^{2+} from gill surface
- leads to additional gill damage
- further increases Na^+ and Cl^- loss
- further inhibits active Na^+ and Cl^- uptake
- increases potential for ionoregulatory toxicity

D. Increasing Aluminium at Moderate pH
- aluminium precipitates on gill surface
- increases mucus production and inflammation
- increases diffusion distances for O_2 and CO_2
- reduces O_2 and CO_2 exchange rates
- increases potential for respiratory toxicity
- allows more toxic Al species to bind to gill
- increases potential for ionoregulatory toxicity

E. Increasing Calcium in Water
- reduces Na^+ and Cl^- loss
- reduces Al binding to gill surface
- reduces ionoregulatory toxicity
- reduces gill damage
- increases diffusion distance
- increases potential for respiratory toxicity

Fig. 8.6—Principal modes of action of aluminium toxicity in fish, alone or in acid and low-calcium conditions (Wood and McDonald, pers. comm.).

where survival $= x$ and concentrations are $\mu eq\, l^{-1}$ for calcium and $\mu g\, l^{-1}$ for aluminium.

Other empirical equations for brown trout have also been formulated for trout density (T) in the field (Ormerod *et al.*, 1990):

$$\log T = -1.24 - 1.08 \log[\text{Al}] + 1.33 \log \text{hardness} - 0.22 \log(\text{flow})$$

where trout density is number per 100 m, [Al] is in mg l^{-1} (filtered), hardness is mg l^{-1} CaCO$_3$, and daily flow is m^3 s^{-1}. Sadler and Lynam (1987) found in sustained laboratory conditions a significant correlation between growth of brown trout, pH and polymeric aluminium (Al*), finding little effect of exposures with pH > 4.3, and for other exposures:

$$\%\text{max. growth rate} = 97.3 - 14.1[\text{Al}^*] - 130.8[\text{Al(OH)}^{2+}]$$

These relationships have been used to predict the status of fisheries in improved water quality following reduction of acid input in rain, or to explain the current fishery status in some acid waters.

8.7　EFFECTS ON INVERTEBRATES AND THE AQUATIC COMMUNITY

The effects of aluminium on other aquatic groups are much less well reported. Some examples from major groups are included here.

Diatoms, algae and macrophytes

The diatom community in acid, high-aluminium, waters is different from that characteristic of circumneutral waters. Reconstruction of past water chemistry from fossil diatom material using correspondence analysis (Birks *et al.*, 1990) relates these differences to aluminium (total), along with pH, calcium, alkalinity and DOC. Experimental exposures at > 800 μg l^{-1} inhibit growth of *Cyclotella* sp. (Rao and Subramian, 1982). Growth rates of *Asterionella ralfsii* were reduced with > 400 μg l^{-1} (Gensemer, 1991), greater at pH 6 where inorganic monomeric aluminium predominates. Similarly, exposures of several algal species to concentrations of about 1000 μg l^{-1} or more were needed to inhibit growth. While *Spirogyra varians* was killed at only 135 μg l^{-1}, a species of *Spirogyra* from a peat bog needed a minimum exposure for 96 h to 27 mg l^{-1} (Bohm-Tuchy, 1960). Macrophytes are less diverse in acid, high-aluminium waters (Stokoe, 1983; Ormerod *et al.*, 1987a), influencing the associated invertebrate fauna. On the other hand, some plant forms, such as liverworts, seem to thrive in these conditions, and to take up and release aluminium in response to short-term changes in water quality (Caines *et al.*, 1985). Changes in the epilithon and epiphyton in a lake acidified to pH 5 over a two-year period showed that the community changed—filamentous cyanobacteria declined and green algae became more abundant (Turner *et al.*, 1991), affecting the rate of carbon fixed by photosynthesis. In a limed lake in the Adirondacks, reverting to acid conditions, chlorophyll decreased but productivity per unit biomass was increased, partly attributed to lower invertebrate grazing, although fewer invertebrates may reflect the presence of fish introduced during the earlier liming phase (Bukaveckas, 1992). Another hypothesis is that aluminium in runoff from acid soils forms insoluble complexes with phosphate within the lake, limiting P uptake by the phytoplankton.

Zooplankton

Among a variety of planktonic crustaceans, daphniids are more sensitive than copepods, and these are more sensitive than dipteran larvae (Brett, 1989). Species sensitivity is widely variable—of ten crustacean species, four showed no reduction of survival at pH 4.5 and aluminium at 500 μg l^{-1} (Havens, 1991). Two of the acid-sensitive species, *Daphnia galeata* and *D. retrocurva*, exhibited aluminium binding at ion exchange sites in the maxillary gland, while a tolerant species, *Bosmina longirostris*, did not (Havens, 1990). Whole-lake acidification or neutralization experiments inevitably lead to changes to the rotifer and crustacean zooplankton communities, but long-term changes are not adequately explained, nor is the role of aluminium.

Molluscs

This group is generally absent or rare in acid, low-calcium, waters, so that toxicity to aluminium is rather irrelevant and few investigations are reported. It appears that molluscs are tolerant of concentrations a 100-times greater than natural concentrations (about 10 mg l^{-1}). Nonetheless, an extensive kill of *Lymnea peregra* was reported during an acid, high-aluminium, episode in a limed Swedish lake (Hultberg and Andersson, 1982).

Littoral communities

Macroinvertebrate communities have characteristically fewer taxa in acid, high-aluminium, waters (Sutcliffe and Hildrew, 1989). Invertebrate responses to acid episodes, however, suggest that both acute mortality and increased drift combine to deplete the exposed community of sensitive species such as the larva of the mayfly, *Baetis rhodani* (Hall *et al.*, 1985; Kullberg and Petersen, 1987; Ormerod *et al.*, 1987b; Raddum and Fjellheim, 1987). However, the relative tolerance of both pH and aluminium of most insect larvae provides a reasonably balanced invertebrate community including stonefly and caddis fly larvae and at least some tolerant mayfly species such as *Leptophlebia spp.*

Amphibia

Many Amphibia utilize temporary ponds for reproduction; these are often acid and low in alkalinity in woodland areas, and most species have a high degree of tolerance to these conditions. A recent review (Freda, 1991) demonstrates that aluminium toxicity to Amphibia is determined, as for fish, by the other water quality components, and that there is wide variation between species and life stages. Embryos are the most sensitive stage to acidity, while newly hatched tadpoles are twice as sensitive to aluminium. With low pH (4.0–5.0) an aluminium concentration of > 200 μg l^{-1} can be lethal. The lower tolerance range of adults is in the pH range 3.5 to 5.0, but embryos (a few sensitive species, see Table 10.4) may be sensitive at pH 3.7 to 4.5 (J. P. Baker *et al.*, 1990). In a field acidification experiment, hatching success of three species was inversely correlated with aluminium and DOC (Clark and Hall, 1985). However, observations show a wide range of effects; there are no studies to date on the effects of acid, high-aluminium conditions on adult populations.

Birds

Some reports of acidification effects on a few bird species feeding in the littoral zone
of lakes and rivers are found in the literature. These seem reasonably attributable to
the reduced feeding potential of the littoral fauna, rather than to aluminium toxicity
per se, or to aluminium accumulation in the food items which is not demonstrated.

8.8   UNRESOLVED ISSUES

The chemical speciation of aluminium compounds in soil and surface waters is now
better documented for stable conditions, but it is still difficult to match with
operational fractions obtained from real samples.

Changes in speciation in the short-term, dynamic conditions in streams or during
high flow conditions are not well established.

In particular, the complexes developed with a variety of inorganic and organic
complexands are poorly understood.

The release of aluminium from soils is still subject to speculation and hypothesis,
although some credible mechanisms have been proposed and could be tested.

The precise toxic species of aluminium present in surface waters are still debated,
but the inorganic monomeric fraction provides a measure of toxic effect.

The same conditions that enhance aluminium in surface waters also lead to a
varied suite of trace metals; their independent or combined effects at realistic
concentrations are not established.

Effects on aluminium availability and speciation by many other chemical components in fresh water as well as temperature, P_{O_2} and P_{CO_2}, are not well documented,
especially their dynamic aspects.

Acute toxic effects of aluminium are well demonstrated for fish, amphibian embryos
and some invertebrates but there are few studies of delayed or insidious effects on
species or on communities.

In many cases the independent effects of aluminium have not been distinguished
from those of acidity or associated conditions.

8.9   SUMMARY AND CONCLUSIONS

Aluminium is ubiquitous and present in all soils, from which it is mobilized to soil
water when the base reserve of soils is exhausted, and where there are accompanying
anions.

These conditions are enhanced by transient acid or salt events in rain, associated
in surface waters with high flows.

On transfer to surface water bodies, dissolved aluminium species may be precipitated
in response to the change in pH or P_{CO_a}. Deposited aluminium on the substrate
materials may be a subsequent source of aluminium if pH falls.

Aluminium (total) concentrations in surface waters are higher with lower pH (i.e.
higher acidity); the same is true of other trace metals (Fe, Mn, Cu, Cd, Zn). This may
not increase their toxicity.

Aluminium complexes with inorganic (silicate, fluoride) or organic (citrate, humates) components in the water leads to reduced toxicity.

Aluminium in the pH range 5–6 is most toxic; it is largely present in hydroxy forms. At lower pH most is present as Al^{3+}, while at $pH > 6.5$, most is present as $Al(OH)_4{}^-$. Inorganic monomeric aluminium is an index of toxic conditions, not total aluminium.

The toxic hydroxy forms are damaging to sensitive fish species at about $60 \mu g \, l^{-1}$, but at even lower concentrations if calcium is $< 2 \, mg \, l^{-1}$. Concentrations as low as $30 \, \mu g \, l^{-1}$ can have long-term effects on fish growth.

The primary effect of aluminium in fish (and some crustaceans) is to impair the gill functions of ion regulation and respiration.

There is little or no accumulation of aluminium in fish or invertebrate tissues; however, aluminium precipitated at the gill surface can cause anoxia as well as adverse morphological changes to the gill surfaces and structure.

Invertebrate and amphibian species are generally less sensitive to aluminium than fish, but there are a few exceptionally sensitive species.

Phytoplankton communities show reduced C fixation in acid, high-aluminium conditions, possibly because of P limitation.

9

Fisheries in acid waters

9.1 INTRODUCTION

There is no doubt that acid waters of pH < 5.0 in temperate regions have fewer species of fish, and often fewer numbers, while pH in the range 6.0–9.0 is usually thought to support good fisheries (Alabaster and Lloyd, 1982). Nonetheless, it should be remembered that some waters of pH < 5.0 do contain healthy fish communities, and conversely, some of higher pH may be fishless. Indeed, the Amazon, with acid tributaries to pH 3.5, has a most diverse fish fauna (at least 1300 species; Sioli, 1975), often utilizing terrestrial fauna or fruit for their nutrition (Henderson and Walker, 1986). Clearly acidity is not the only criterion for fish absence or loss; other toxic agents may be present and other natural conditions (e.g. altitude, temperature, flow, lake and stream morphology) may limit species diversity.

This chapter will review evidence for a relationship between acidity (and related water chemistry) and fish presence, diversity, abundance and well-being. It will also consider the evidence for a trend in fishery status in relation to the record of acidification of surface waters. Other biological components of freshwater ecosystems, including plankton, macroinvertebrates, amphibians, and riparian birds and mammals, will be considered in Chapter 10. It should not be forgotten that a balanced community structure is necessary for a successful fishery.

9.2 FISHERY STATUS OF ACID WATERS

It has often been claimed that acid waters are 'dead', an extreme and untenable view in waters of pH > 3.0! It is generally accepted, however, that waters of pH < 5 are unfavourable to fish, and that their variety, abundance and growth are reduced in acid compared with circumneutral waters. Several studies of fishery status of acid waters are available as well as more intensive studies of the fish communities of some lakes and river systems. These are principally for a number of acid waters in Scandinavia and North America (Table 9.1 gives some regional examples). No

Table 9.1—Sources of synoptic fisheries survey data, relevant to acid
conditions

Source reference	Region	Number of lakes	Number of species
Rahel, 1982	Northern Wisconsin, USA	138	48
Harvey & Lee, 1982	La Cloche, Ontario	63	32
Kretser *et al.*, 1989	Adirondacks, New York	1469	53
Baker *et al.*, 1990	US northeast states	71	18
Matuszek & Begg, 1986	Ontario, Canada	2931	—
Keller, 1984	Florida	49	12
Kentammies, 1991	Finland	780	
Almer *et al.*, 1974	West coast, Sweden	50	6
Almer & Hanson, 1980	West coast, Sweden	55	17
Dickson, 1975	Lakes south of Dalalven, C. Sweden	36	11
Degerman & Sers, 1992	All Sweden	981	37
Henriksen *et al.*, 1988b	Norway	1005	> 13
Maitland *et al.*, 1987	Scotland	83	10

comparable studies are available for the UK, partly because of the relative paucity of species (*cf.* 55 UK species but 80 in northwest Europe and 215 in Europe as a whole) and partly because there are almost no lakes or streams in the UK unaffected by human interference. In some regional surveys there is evidence, albeit anecdotal, of a change of fish species with progressive development of acid conditions (Fig. 9.1).

On the other hand, there are evidently fishless waters where other explanations are just as credible—for example, where stock has been killed by pollution and has not been able to recolonize, where habitat is unsuitable or has been changed by water management practice, or where spawning conditions are unfavourable. Streams are much more at risk from short-term conditions, including acid episodes as well as hydrological events, both liable to displace fish or spawning grounds. These events apply to circumneutral as well as acid waters (see Chapter 6). In contrast, there are waters of significant acidity where some fish are found unexpectedly, possibly due to physiological acclimation, selection of tolerant strains, or in refuges from the main stream conditions.

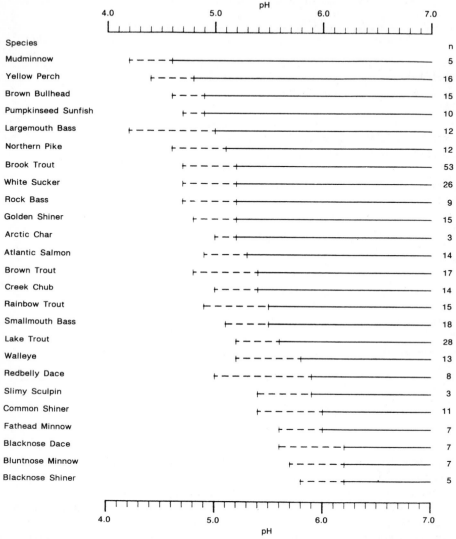

Fig. 9.1—Distribution of some European and North American fish species in relation to surface water pH; estimated critical values for fish population effects, with dashed lines indicating uncertainty (after J.P. Baker *et al.*, 1990). (n is the number of observations used to derive the critical pH.)

9.3   REGIONAL FISHERY SURVEYS

Sweden

Early studies in Sweden (Almer *et al.*, 1974; Dickson, 1975; Fig. 9.2) documented the loss of roach (*Rutilus rutilus*), perch (*Perca fluviatilis*), pike (*Esox lucius*) and char

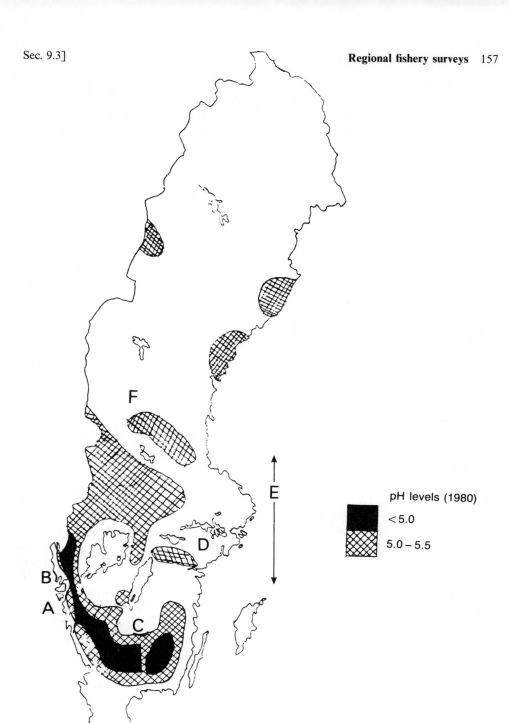

Fig. 9.2—Fishery status reported in Sweden. **A**: West coast lakes and streams; **B**: Bohuslan lakes; **C**: Småland lakes; **D**: Eastern Swedish lakes; **E**: Dickson's lakes; **F**: Central Swedish lakes.

(*Salvelinus alpinus*) from several lakes in south–central Sweden, contemporary with acidification of some lakes in that area from the 1930s to the 1970s. The lakes which lost char were at pH < 5.0 and zero alkalinity, but four lakes in a 'control' set of 29 with pH > 5.0 also lacked char. In 50 west coast lakes roach were also absent in 10 of 13 lakes with pH < 5.0 and perch and pike were reported 'disappearing'; six of 37 west coast lakes with pH > 5 also had no roach (Almer *et al.*, 1974). In a sample of 100 lakes with pH 4.4–7.5, test fishings found the minnow (*Phoxinus phoxinus*) in only 16 lakes; this species was considered the most sensitive on this basis (Almer *et al.*, 1978). Whitefish (*Coregonus albula*) was present in 19 lakes and trout (*Salmo trutta*) and char were found in 25 and 24 lakes respectively. Perch were found in 95 lakes. Only two of the 100 lakes, however, with pH values of 4.5 and 4.65, had no fish.

In a wider-ranging survey of 1700 lakes north of Gothenburg, a third had pH < 5.3 on the single sampling occasion, but perch were present in most and char and trout absent from only one lake, while roach were absent from several (Schmuul, 1976). This and similar water quality surveys in other regions in Sweden led to the estimate that 18 000 (20%) of the total of 90 000 lakes throughout the country, as well as 90 000 km of running waters, have a pH regime which has damaged or threatens fisheries and other aquatic life (Swedish Min. Agric., 1982). Acidification in Sweden is generally thought to have occurred in the past few decades (since the 1940s), but some of the fish losses recorded seem to have predated this. Resampling of the 50 lakes in south–central Sweden in 1980 found additional roach and whitefish populations, and recovery of some perch populations, some because of liming or documented improvement in water quality, although there has been little general change in surface water quality following the reduction in sulphur emissions since the late 1960s (Sanden *et al.*, 1987).

Fish communities in nearly 1000 stream reaches throughout Sweden were sampled by electrofishing by Degerman and Sers (1992). Streams have relatively few fish species (13 caught frequently) and their communities are limited by five components that explain 78% of the variation: climate, colonization potential, stream order, depth, and distance from the nearest lake. Salmonid species were more frequent in the extreme, low-competitive, environments. Water quality seems to be a less important factor and Henderson (1985) distinguished fish communities in Swedish lakes (as reported by Almer and Hanson, 1980) as a cyprinid set limited to waters of pH > 5.0, a set including pike, perch and eel found in most waters, independently of pH, and a third, only brown trout, limited to waters above 100 m altitude with pH > 4.5 (Table 9.2).

Norway
A decline since about 1915 in the yield of Atlantic salmon fisheries for seven southern rivers of Norway has been compared with that of less acid rivers to the north (Muniz, 1981); these southern rivers were relatively unproductive even at the turn of the century. Loss of brown trout populations in southern lakes since the 1940s was about 50% by the 1970s (Sevaldrud *et al.*, 1980), as estimated from angler questionnaires. A further loss of 30% of brown trout and 12% of perch populations is reported since

Table 9.2—Environmental requirements for some European and North American fish species

Fish species	Environmental needs
Eel (*Anguilla anguilla*)	Present in most waters to pH 4.0
European perch (*Perca fluviatilis*)	Present in most waters, fry tolerant at pH 4.0
Yellow perch (*P. flavescens*)	Ca > 4 mg l^{-1}, pH > 4.2
Brook trout (*Salvelinus fontinalis*)	pH > 4.2
Pumpkinseed (*Lepomis gibbosus*)	Ca > 4 mg l^{-1}, pH > 4.5
Brown trout (*Salmo trutta*)	pH > 4.5, alt. > 100 m
Lake trout (*S. namaycush*)	pH > 4.5
Pike (*Esox lucius*)	Present in most waters to pH 4.5, lake size
Northern pike (*E. lucius*)	Ca > 5 mg l^{-1}, lake size > 50 ha
Whitefish (*Coregonus albula*)	pH > 4.5
Lake herring (*C. artedii*)	Max. lake depth > 5 m
White sucker (*Catastomus commersoni*)	pH > 4.8, inlet/outlet streams
Lake whitefish (*C. clupeaformis*)	pH > 5.0, lake size
Rudd (*Scardinius erythrophthalmus*)	pH > 5.0
Tench (*Tinca tinca*)	pH > 5.0
Salmon (*Salmo salar*)	pH > 5.0
Walleye (*Stizostedion vitreum*)	pH > 5.2, lake > 190 ha
Rainbow trout (*Oncorhynchus mykiss*)	pH > 5.5
Bullhead (*Ictalurus nebulosus*)	Ca > 5 mg l^{-1}, lake > 80 ha
Minnow (*Pimephales promelas*)	pH > 5.8
Johnny darter (*Etheostoma nidra*)	pH > 5.9
Iowa darter (*E. iowae*)	pH > 5.8
Bluegill (*Lepomis macrochirus*)	Inlet/outlet streams

then (Sevaldrud and Skogheim, 1986). Several surveys have been undertaken in the area, the first (700 lakes) in 1974–75 by Wright and Snekvik (1978), followed by that of Sevaldrud *et al.* (1980: 1200 lakes), both based on catch returns, and recently by Henriksen *et al.* (1987, 1988) reporting on both chemistry and fishery status in the Norwegian '1000 Lakes Survey'. Some of the earlier Wright and Snekvik lakes were resampled in 1986, providing an assessment of the water quality and changes in several fish species over more than a decade (Fig. 9.3).

Wright and Snekvik provided a systematic database linking water chemistry and reported fishery status; the lakes were predominantly salmonid and perch fisheries. Although fish loss continued in the ten years 1976–86, there were few changes in water chemistry—acidity and sulphate were much as earlier but calcium and magnesium fell. Both low calcium and high acidity are well correlated with fish

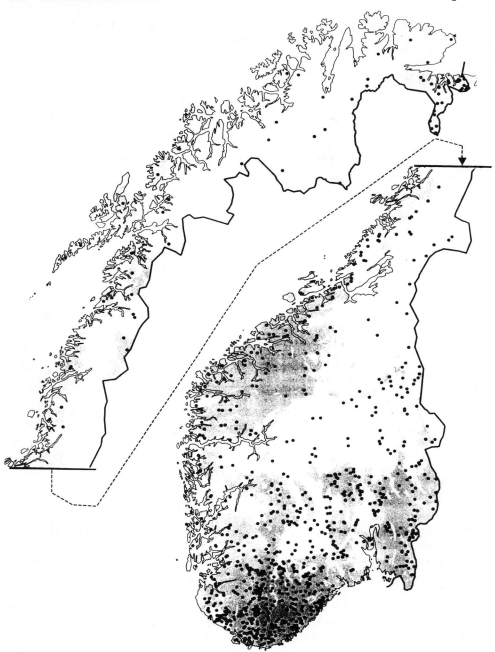

Fig. 9.3—Assessment of fishery status of (southern) Norwegian lakes in relation to water quality (after Henriksen *et al.*, 1988b). The areas underlain by granite are shaded; most fishless lakes are in the southernmost area.

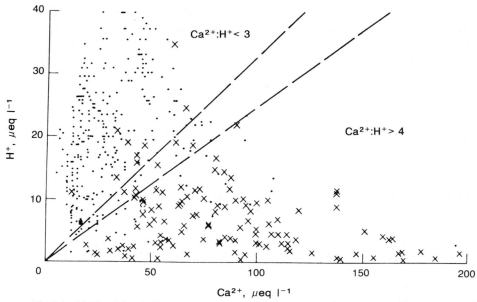

Fig. 9.4—Fishless lakes (●) in Norway compared with lakes with good populations (×), in relation to the levels of hydrogen ion [H^+] and calcium [Ca^+] in the lake waters (after Chester, 1984).

absence; the ratio of [Ca^{2+}] to [H^+] appears to be critical, a value of less than 3 being seen in almost all the fishless lakes (Chester, 1984; Fig. 9.4). While a statistical relationship can be found for fishery status and pH, an equally convincing one is found with calcium. However, the field data for southern Norway have been interpreted using multivariate analysis to show that low calcium is not harmful (Muniz and Walloe, 1990). This interpretation is challenged on the basis that many water quality components are dependent and in oligotrophic upland lakes of Norway calcium is in the critical range of concentration; Rosenberg (1990) argued that trout stock strength is best explained statistically by pH, aluminium and altitude, and that calcium is likely to be inversely related to altitude and a linear relationship (Ca versus fishery status) may not be appropriate. The role of calcium has since been confirmed overwhelmingly for a wide variety of fish species in a range of pH and calcium conditions in both field conditions and in laboratory tests (J. P. Baker *et al.*, 1990).

The abundance (as catch per unit effort) and population structure of perch populations in 30 acid lakes in Norway (pH 4.3–5.9; Ca 0.4–2.4 mgl^{-1}; TOC to 14 mgl^{-1}) is reported (Hesthagen *et al.*, 1992). Some populations have declined, mostly due to recruitment failure, but associated with higher TOC and calcium, rather than low pH or high inorganic aluminium.

Finland
The majority of Finnish lakes (56 000 of > 1 ha) have acidity dominated by organic anions; sulphate deposition has raised lake water sulphate concentrations in the

south, where many lakes have pH < 5.0. However, in 1985, the HAPRO project (1985) reported that 'fish population loss has not been found for a single lake'. A few years later, sampling in 987 lakes found that roach stocks were reduced in at least 780 lakes south of 66°N, and this species had been lost in 330 lakes (Kentammies, 1991); similarly, perch had been affected in 110 lakes and lost in 12 lakes. In two acid lakes in southern Finland, the density of perch populations has fallen in the past decade, attributed to failure of reproduction. However, some fry were recruited and grew faster than the earlier, more abundant, year classes, an evident density-dependent response (Rask, 1992).

United Kingdom
Similar surveys have not been undertaken in the UK, although some data exist for Scotland (Maitland *et al.*, 1987). Records of catches of migratory salmonids in rivers have fallen over the past few decades, following anomalously high catches in 1955–65, and have not recovered. The rivers most affected were in the west and north, where surface water pH is lower than in eastern rivers. Over the longer term, Scottish salmon catches have fluctuated widely over the past 100 years, with numbers of spring-returning salmon declining recently in favour of a greater run of grilse (one sea-winter fish). It has to be recognized that there are few quantitative data for migratory salmonids, with their complex life history and changing patterns of management and exploitation. A clear relationship linking the decline of salmon (but not grilse) with increasing extent of conifer afforestation in western Scotland has been reported (Egglishaw *et al.*, 1986). Similar independent reports for Wales (Stoner and Gee, 1985; Warren *et al.*, 1988), based on rod catches from 22 rivers, are linked to the extent of acid catchment areas; however, a parallel decline in net catches is independent of acid runoff. It is difficult to assess the relationship of these changes with stream conditions alone since there have been significant changes in the level of exploitation (including sea catches), the incidence of disease, e.g. UDN (ulcerative dermal necrosis) in the 1960s, and of upland land management. The latter includes a reduction in upland liming, an increased use of agricultural chemicals, and an increase in organic pollution incidents. However, changes in upland reaches where salmonids spawn would be expected to affect recruitment and rod (river) and net (estuary) catches alike, and of both salmon and sea trout (Jenkins and Shearer, 1986).

A parallel assessment of the status of brown trout in UK waters is not possible as systematic records are not kept. In the Galloway area (southwest Scotland), however, there are historic records of annual trout catches from the 1940s (Harriman *et al.*, 1987) for several lakes. In two (Fleet and Grannoch), where pH values were 4.3 and 4.6 respectively in 1978/79, the trout catch fell to zero; paradoxically, the Fleet catch fell in 1955 although circumneutral pH values in the lake were reported through to 1970. Twelve lakes draining granite and peaty terrain in the same area have pH < 4.8, of which six are fishless; two have no earlier records of fish.

Investigation of the fish populations of 83 acid-sensitive upland lakes and their tributary streams in Scotland has shown that they have a restricted variety of species; 14 are reported to be fishless, although most have some record of trout in the past. The fishless lakes have low calcium concentrations $(1-2\ mg\ l^{-1})$ and low pH (< 5.2)

and almost all are in the high-rainfall area in the southwest (Galloway, Islay, Arran) draining granite and peaty catchments; some are very small (Maitland *et al.*, 1987). They are also affected significantly by 'sea salt rains' which mobilize aluminium from acid soils (see Chapter 7). Elsewhere in Scotland other lakes are also acid and low in calcium, but only four are fishless.

In mid and west Wales lakes draining catchments with shales and mudstones, and in the north draining granite, are acid with low calcium concentrations. These lakes have fewer trout than those which are less acid in the same regions; anglers' catch records indicate that trout catches have declined in the last 20 years (although angler numbers have increased) (Stoner and Gee, 1985); trout are often absent in the more acid streams and restocking has failed. Land use and management have changed in the area, with extensive afforestation. In an independent survey of streams in this area, along with others in the Pennines, 45 streams of 61 sampled showed a positive correlation of fishery status and water quality, particularly with calcium and alkalinity, and an inverse relationship with acidity, aluminium and heavy metals (Turnpenny *et al.*, 1987). Sixteen of the streams were fishless even though of circumneutral pH, possibly because of the combined toxicity of heavy metals in some, or because of recruitment risks in first-order streams since $0+$ and $1+$ group fish were absent in more than half. Species other than trout (eels, bullheads, minnow) were also restricted or absent.

In Cumbria (northwest England), upper reaches of two acidic rivers (Esk and Duddon) draining a catchment of volcanic geology are fishless or have a low population density (Prigg, 1983), although this may not be a new phenomenon; at the end of the last century an angling author (Watson, 1899) noted that the area was more noted for its scenic beauty and its poetry (*vide* Wordsworth) than the abundance of its fish. Fish kills in the lower reaches of the Esk have also been reported following heavy rain.

The reported distribution of fishless waters in the UK is shown in Fig. 9.5; they are predominantly small lakes in heavy rainfall areas where geology and soils indicate susceptibility to acidification (Warren *et al.*, 1986, 1988).

United States

Lakes and ponds in the Adirondacks (upper New York State) have been studied since the 1920s for water quality and fishery status. Recent assessment of this database (J. P. Baker *et al.*, 1990) has provided insight into this relationship and the causes of fishery changes. A state survey in the 1920s and 1930s provides a baseline, although earlier methods of chemical analysis are not compatible with modern records (see Chapter 6). With growing interest in acid waters in the 1970s, a study of 217 lakes above 600 m was undertaken (Schofield, 1976). Of 40 lakes, 19 were fishless and had pH $<$ 5.0; later analysis (J. P. Baker *et al.*, 1990) found that three were fishless in the 1920s, and another three had lost their fish by 1990.

Notwithstanding the bias introduced by changing chemical methods, a significant shift in pH of 0.12 pH units per year was estimated up to 1979, and associated with 17 lakes becoming fishless over that period. The most recent analysis (including a dataset of 1469 lakes $>$ 0.5 ha) found that 219 lakes have pH $<$ 5.0, of which about

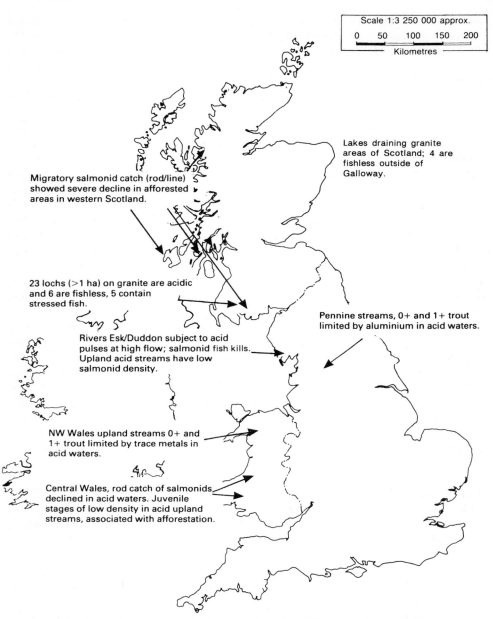

Scale 1:3 250 000 approx.

0   50   100   150   200
Kilometres

Lakes draining granite
areas of Scotland; 4 are
fishless outside of
Galloway.

Migratory salmonid catch (rod/line)
showed severe decline in afforested
areas in western Scotland.

23 lochs (>1 ha) on granite are acidic
and 6 are fishless, 5 contain
stressed fish.

Rivers Esk/Duddon subject to acid
pulses at high flow; salmonid fish kills.
Upland acid streams have low
salmonid density.

Pennine streams, 0+ and 1+ trout
limited by aluminium in acid waters.

NW Wales upland streams 0+ and
1+ trout limited by trace metals in
acid waters.

Central Wales, rod catch of salmonids
declined in acid waters. Juvenile
stages of low density in acid upland
streams, associated with afforestation.

Fig. 9.5—Fishery status studies in acid waters in the UK (after various authors).

half have acidity attributable to organic anions; 125 lakes have mineral acidity, of
which 31 are < 4 ha and above 600 m. On the basis of observations elsewhere, it
was judged that 12 would be naturally fishless. While the fishless status of about a

third of the lakes is thus reasonably attributed to acid rain, it was concluded that brook trout populations were influenced more by biotic than chemical interactions with the ranking:

$$\text{stocking} > \underset{\text{species}}{\text{associated}} > \underset{\text{species}}{\text{competitive}} > \underset{\text{pCa:pH}}{\text{pH,}} > SiO_2 > ANC$$

The limiting water quality conditions for most fish species present are pH, aluminium (monomeric inorganic) and calcium, with silicate (indicating a groundwater influence) important for brook trout; over 50 significant variables were identified but many were eliminated by collinearity.

In Adirondack waters (Kretser *et al.*, 1990), 53 species of fish are recorded, with a variety of salmonids, cyprinids and several minnow species sensitive to pH < 5.0. Four species (central mudminnow, brown bullhead, golden shiner and yellow perch) were found in lakes of pH < 4.5 and an additional 11 species were present at pH < 5.0. Fish were absent, however, in 346 lakes (24% of the sample), representing 7% of the lake area. Although brook trout were the focus of investigation and were widely impacted, they can live in a range of pH 4.5–9.0 and it was concluded that 'acidification was not the only factor, or necessarily the dominant factor, influencing Adirondack fish communities'. Alternative explanations of the changes were the 'reclamation' with rotenone of some lakes for sport fish restocking, changes in stocking policies, and the introduction of competitive or predatory species.

Streams provided limited data, but confirmed the conclusions made for lakes; no fish were found in streams with pH < 5.2.

In other northeast regions (Fig. 9.6), surface waters show much less 'acidic stress' (defined by water quality) by a factor of 2 to 3—about 5% of waters in the northeast areas were judged unsuitable for brook trout, and 9% unsuitable for acid-sensitive species. In Maine, no clear association was found between fish population losses and acid–base chemistry of lakes (Haines, 1987). In Vermont and New Hampshire (Langdon, 1985), trends suggest a pattern similar to the Adirondack lakes—some of the acid fishless lakes have always been acid, while others support natural and stocked populations. Mid-Appalachian streams (J. P. Baker *et al.*, 1990) have examples of fish kills and loss of some populations with increasing acidity, but systematic survey data are lacking. About 18% of streams are unsuitable for brook trout survival, and about 30% unsuitable for more sensitive species. In the upper Midwest, few fishless lakes were found, although the more acid waters had fewer fish species. In contrast, fish species common to Florida lakes have reproducing populations, e.g. largemouth bass, even at very low pH (4.0–4.5) and there are no evident population losses.

Canada

Acid lakes close to the giant metal smelter at Sudbury, Ontario were highly acid and also contaminated by large heavy-metal concentrations. These lakes suffered massive near-field deposition of both sulphur and metals, as well as acid rain, for many decades. Some of the lakes lost their fish populations, but the presence of high levels of toxic metals, their high conductivity and total solutes, including base cations, makes them quite different from waters acidified by acid rain. At some distance

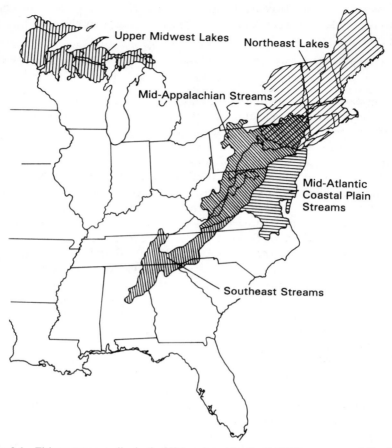

Fig. 9.6—Fishery status studies in the USA reviewed by the NAPAP programme (after J.P. Baker *et al.* 1990).

downwind in the La Cloche area where lakes drain forested quartzite catchments, conditions are more comparable with those in acidified areas of northern Europe. Fish populations in these lakes have been lost in 50 lakes of pH < 6.0 (Harvey and Lee, 1982); the number of fish species was related to lake pH and distance from the Sudbury source, but only 5% of lakes in southeast Ontario have zero alkalinity (see Chapter 7). Some are also contaminated with heavy metals (Beamish, 1976). Of 31 lakes with pH < 5.0, 14 have no fish. Fish population losses were initially of most sensitive species, with reproductive failure in those that survived (Magnuson *et al.*, 1984). Brook trout were re-established in one lake when pH rose above 5.0 (Beggs and Gunn, 1986), and in another lake, trout reproduction resumed at pH > 5.5 (Gunn and Keller, 1990); one consequence, however, was a sharp decline in perch due to trout predation, and minnows increased but white suckers and burbot did not recover (Keller *et al.*, 1992).

In Ontario, about 200 lakes are considered to have lost fish populations, 104 of them lake-trout lakes; field surveys in nearly 3000 lakes suggest that several species are tolerant of low pH (yellow perch, red-belly dace, pumpkinseeds); their absence probably relates to zoogeographic factors rather than water quality. Other species, however, are sensitive, including small cyprinid minnows, as in Adirondack waters. For larger fish such as lake trout, northern pike, walleye, whitefish, and shiners, frequency of occurrence is related progressively as pH falls below 6.4. Although lake trout are limited in the low pH lakes, 55% of the Ontario stock is found in areas receiving high acid deposition and where the lakes have low alkalinity (Wales and Beggs, 1986). In this region, reduced deposition since 1980 has now led to the return of some species, mostly in the rank order of their sensitivity to acidity.

In Nova Scotia, spawning Atlantic salmon have been lost from 10 rivers in the southern uplands, with a present pH < 5.0, while in 12 other rivers with higher pH, they are not affected (Watt *et al.*, 1983). In these rivers, acidity rather than toxic aluminium is identified as the cause, since most aluminium is complexed with organic material (Lacroix and Kan, 1986; see Chapter 8). The estimated area lost to salmon spawning in Nova Scotia is, however, only 6 km^2 (16% of the total) although a larger area (13 km^2) is considered as threatened since alkalinity is $< 75 \mu$eq l^{-1} (Stewart *et al.*, 1988). In Labrador and Quebec, regions of similarly poor geology and with low-alkalinity waters, there is no record of lost salmon spawning areas and over a nine-year period there was no relationship of fish catch with lake acidity in 158 Quebec lakes (Richard, 1982).

9.4 FISHERIES AND WATER QUALITY

Biological observations of damaged fish communities in acid waters show that the number of species is less than expected, that the structure of populations is anomalous, and productivity and yield are reduced.

Species richness

Loss of some fish species with increasing acidity results in a decline in **species richness**; in several regional surveys the number of species present is less in waters of low pH, although there are differences between regions (J. P. Baker *et al.*, 1990). Significantly fewer species than expected were found in acid lakes and streams (pH < 6.0) in the Adirondacks (New York), northern Wisconsin, and Quebec (J. P. Baker *et al.*, 1990). However, for Maine lakes and Vermont streams no clear relationship was found (Haines *et al.*, 1986; Langdon, 1985). In the Netherlands, a survey of 91 soft-water lakes found only two species in lakes of pH < 4.0, eight species in lakes of pH 4.0–5.0 and 20 species in lakes of pH > 5.0 (Leuven *et al.*, 1987). Comparing the fish communities of acid La Cloche (Ontario) lakes and Swedish lakes, Henderson (1985) suggested that fish assemblages are predictable on the basis of a limited number of environmental requirements (Table 9.2): pH, calcium, lake size and depth, inlet and outlet streams and altitude were identified as important, and in some cases also biotic competition. For other data sets from Wisconsin and from other Ontario regions, species interactions often seem to limit species frequency.

Recruitment

In a healthy population of fish, numbers in the first year, the **recruitment class**, are usually high and most natural losses occur during this early stage of the fish life-history; alevins and fry are particularly at risk. In following years survival is generally good, and each year class should be represented in the population, possibly even post-spawners, for example for brown trout. In acidic waters, in contrast, some classes may be entirely absent, or grossly depleted, possibly because of conditions occurring during the more sensitive juvenile stages, but also because recruitment has fallen below that needed to maintain the population (McFadden, 1977). Episodic and extreme events which occur in sensitive upland streams, even in areas of minimal acid deposition (see Chapter 6), provide a reasonable explanation. The gapped year classes are evidence, often confirmed by chemical records, that transient adverse conditions occurred, perhaps as a result of snowmelt during fry emergence. Recruitment failure is perhaps the most common cause of fish population decline (Rosseland, 1986).

Another explanation of lack of recruitment may be that egg production is affected by exposure to maintained low pH and associated conditions (e.g. for perch: Hesthagen et al., 1992; Rask, 1992); this was seen for some species in the experimental lake acidification of Lake 223 (Mills et al., 1987) but not for lake trout in acid Ontario lakes (Liimatainen et al., 1987). In general, fecundity in acid waters is equal to, or even greater than, that in circumneutral waters (J. P. Baker et al., 1990).

It is often reported also that fish populations in acid waters have few older and larger fish, possibly the remnants of pre-acid populations, or possibly due to the lack of competition as the lake acidifies, or to reduced angling pressure. This pattern was observed for a variety of species in Ontario lakes (Harvey, 1978), for char and trout in Norway (Muniz and Leivestad, 1980; Andersen et al., 1984) and for perch in Finland (Lappalainen et al., 1988). In contrast, missing older classes and populations dominated by young fish have been found in some acid waters in Norway (Rosseland et al., 1980; Andersen et al., 1984) and Ontario (Trippel and Harvey, 1987). The reasons for such a variety of population responses are not clear.

Biomass production and yield

Fishing success for brook trout in Quebec lakes was related to fish size, lake area and elevation and lake productivity but was not directly related to lake pH: a better relationship was found with calcium, possibly an index of lake productivity (Frenette et al., 1986). In cases where biomass, production or yield is apparently lower in acid waters, it is not clear whether the populations are in equilibrium or in transition, nor is the relative role of acidity or other indices of productivity, usually low in oligotrophic acid-sensitive lakes.

Although some fish prey items may be absent from low pH waters, there is seldom evidence that fish become starved; indeed stomach fullness, growth rates and condition are comparable or even exceed those of fish in circumneutral waters (J. P. Baker et al., 1990). Again, some investigators have reported no clear relationship between pH and fish growth in the field, or slower growth, although this was found in long-term

Table 9.3—EIFAC pH criteria for the protection of European freshwater
fish (after Alabaster and Lloyd, 1982)

3.0–3.5	Unlikely that any fish can survive for more than a few hours in this range although some plants and invertebrates can be found at pH values lower than this.
3.5–4.0	This range is lethal to salmonids. There is evidence that roach, tench, perch and pike can survive in this range, presumably after a period of acclimation to slightly higher, non-lethal levels, but the lower end of this range may still be lethal for roach.
4.0–4.5	Likely to be harmful to salmonids, tench, bream, roach, goldfish and common carp which have not been previously acclimated to low pH values, although the resistance to this pH range increases with the size and age of the fish. Fish can become acclimated to these levels, but of perch, bream, roach and pike, only the last named may be able to breed.
4.5–5.0	Likely to be harmful to the eggs and fry of salmonids, and adults, particularly in soft water containing low concentrations of calcium, sodium and chloride. Can be harmful to common carp.
5.0–6.0	Unlikely to be harmful to any species unless either the concentration of free carbon dioxide is greater than 20 mg l^{-1} or the water contains iron salts that are freshly precipitated as ferric hydroxide, the precise toxicity of which is not known. The lower end of this range may be harmful to non-acclimated salmonids if the calcium, sodium and chloride concentrations, or the temperature of the water, are low, and may be detrimental to roach reproduction.
6.0–6.5	Unlikely to be harmful to fish unless free carbon dioxide is present in excess of 100 mg l^{-1}.
6.5–9.0	Harmless to fish, although the toxicity of other poisons may be affected by changes within this range.

laboratory acid exposure, including associated levels of calcium and aluminium, of
brown trout (Sadler and Lynam, 1986, 1987, 1988).

9.5 MECHANISMS OF ACID TOXICITY

Other manifestations of exposure to acid conditions both in the field and supported
by laboratory studies include physiological distress, behavioural anomalies, poor
growth and skeletal calcification. These effects are attributed to acidity although
associated water quality (calcium or 'hardness' or trace metals) are also involved.
Critical conditions for acidity (pH) in natural waters are moderated by calcium or
'hardness' (Table 9.3) which limits gill membrane permeability. Relative toxicities of
many trace metals in acid waters (Table 9.4) are also moderated by hardness
(Alabaster and Lloyd, 1982; Howells, 1994). Although alkalinity is considered valuable
as an index of fish production in warm-water lakes, this aspect has seldom been

Table 9.4—The relative toxicity of some trace metals in
acid waters (after McDonald *et al.*, 1989)

Trace metal	Upper concentration (μmol)	LC_{50}
Copper	0.3	0.3
Cadmium	0.01	0.3
Zinc	1.9	1.4
Lead	0.3	5.6
Aluminium	36	7.3
Iron	40	7.3
Manganese	10	290
Nickel	0.9	338
Protons (H^+)	100	100 (pH 4)

investigated for acid oligotrophic lakes. Rather, it is the level of calcium (associated with carbonate, it also contributes alkalinity) which appears to be critical. 'Hard' waters have typically 500–5000 μeq l^{-1} (10–100 mg l^{-1}) of calcium, while 'soft' waters have typically only about 100 μeq l^{-1} or less of calcium. Acid waters in Norway often have calcium concentrations < 50 μeq l^{-1}, or a 'hardness' of about 100 μeq l^{-1}. North American waters have higher concentrations of both calcium and alkalinity. Whereas alkalinity reflects the balance of acid and base components of water chemistry (see Chapter 6), calcium specifically affects membrane function, especially that controlling ion regulation in fish (Wood, 1989; see Fig. 8.5). Tolerance to acidic conditions in fish and other aquatic organisms seems to be determined by calcium as much as hydrogen ion concentrations, controlling the ability of the gills to maintain a healthy ion balance between the ambient medium and the blood and tissues. Aluminium, present in relatively high concentrations in acid waters, is also an important toxic agent *per se* (see Chapter 8).

The primary mechanism for the toxicity of acid surface waters is the disruption of normal ion regulation at the gill (McWilliams and Potts, 1978; Wood and McDonald, 1987); exposure to low pH and/or elevated aluminium causes the loss of plasma electrolytes due to both increased rates of loss of ions, and decreased rates of uptake (see Chapter 8). If calcium is low in the ambient medium, the gill becomes more permeable, enhancing the rate of ion loss and increasing sensitivity to pH and aluminium. In some conditions, respiratory impairment occurs (Wood *et al.*, 1988), leading to acidosis and blood changes (haematocrit, haemoglobin, RBC and reticulocytes all increase) (Wood, 1989). In addition, there are changes in gill histology with acid/aluminium exposure consistent with ionic and respiratory stress, including increase in the numbers and density of chloride cells, hyperplasia of the epithelium of the primary lamellae, and increased diffusion distance (Jagoe *et al.*, 1987; Leino *et al.*, 1987). Acid/aluminium exposure also affects calcium metabolism, with lower accumulation and allocation to skeleton, leading to delayed skeletal development in trout (Reader *et al.*, 1988) and skeletal deformities in white sucker (Fraser and Harvey,

1982). During development, acidic conditions also reduce the activity of the hatching enzyme, chorionase, in some cases reducing hatching success when the emerging embryos are not able to break free from their egg capsules (Runn et al., 1977; Peterson et al., 1980).

Bioassay experiments with fish of different ages and species exposed to various pH, calcium and aluminium levels have provided a quantified interpretation of the role of these three variables. (For reviews of a range of sensitive to tolerant species, see J. P. Baker et al., 1990, and Chapter 8.) For brown trout in low pH (< 4.5) waters, fish absence can be attributed to low calcium ($< 50 \, \mu eq \, l^{-1}$) and total aluminium in excess of $100 \, \mu g \, l^{-1}$. In waters of more moderate acidity (pH 5.0–6.0) but low calcium, toxic aluminium species can be held responsible for fisheries damage, especially in waters subject to extreme flows associated with snowmelt or heavy rains. Other modifying agents (citrate, silicate, fluoride, humics) of aluminium toxicity have been discussed in the previous chapter; only humic materials appear to be effective in the field.

Different species have evidently different levels of tolerance and some life stages are more susceptible. Newly hatched fry, with greater ion demand during growth, immature membranes and uncertain homeostasis function, are more sensitive to acid, low-calcium conditions. In contrast, larger fish, with greater respiratory demand, seem more sensitive to aluminium toxicity. Spawning adult salmon or downstream-migrating smolts also have transient demands on their ion regulatory abilities, and are then particularly sensitive to acid and associated water quality conditions (Potts et al., 1990).

It is clear from the foregoing text that water quality, including acidity, is crucial to the success of a fishery. Nevertheless, there are many other factors to take account of—the suitability and extent of the habitat available, especially for spawning, the supply of food items, and interactions with competitive or predatory species. These are independent of water quality and involve population or community interactions. The strength of a population is determined by the balance of recruitment and loss; many biological factors enhance or reduce recruitment or mortality, especially density-dependent functions such as fecundity, spawning success, fry survival and growth, often responding to maintain the 'stability' of the population, a compensatory function which allows exploitation while sustaining yield (McFadden, 1977). It should also be accepted that extreme physical or chemical events unrelated to acidification may be responsible for fish losses or declines. Many acid waters are low-order streams draining uplands with adverse physical conditions associated with variable flow regimes, which may exacerbate acid water quality.

9.6   UNRESOLVED ISSUES

Fish populations are determined by a wide range of physical, chemical and biological conditions. Although water quality plays a part, it is not always clear that this is always the determining factor in fishery status of acid waters.

Regional studies of fisheries status in susceptible waters provide evidence that water quality is paramount in some, but not all, areas. Anomalous features, possibly reflecting short-term events, different species or strains, or inadequate data still await explanation.

Specific criteria for acid exposure are difficult to establish; response varies for species, strains, life stages, and previous exposure, as well as associated water-quality components and independent biological interactions.

The relative importance of water-quality components countering acid or aluminium toxicity (silicate, fluoride, organic complexing agents) is not well established in the field.

Short-term acid events leading to fish kills are demonstrated, but are these sufficient to explain population loss which occurs also in pristine areas with low acid deposition?

Do successive short-term events acclimatize fish effectively to later or sustained conditions?

9.7   SUMMARY AND CONCLUSIONS

Fish communities in acid waters are restricted to species tolerant of low pH and associated water quality. Species known to be sensitive to pH < 5.0 are salmonids, cyprinid minnows, white suckers, and roach fry.

Other species are much less sensitive—in particular, eel, whitefish, pike and perch are often found in waters of pH < 4.5.

Within species, different strains may show more or less tolerance to acidity, because of either acclimation or genetic selection. Different life stages also have different tolerance; the early fry and migrating stages of almonids are more sensitive than other stages and recruitment may suffer.

In northern Europe and North America, acid water conditions have restricted some populations, and some waters of pH < 5.0 are fishless.

10

Effects on other aquatic biology

10.1 INTRODUCTION

Fisheries are the major natural resource of economic value affected by acidification, but there is also concern regarding other components of the aquatic community, including riparian species that are dependent on the proper functioning of the aquatic ecosystem. Particular concern is expressed as to whether certain 'key' species are present, or whether rare species are absent. Are essential biological processes of the ecosystem, such as photosynthetic production, degradation and nitrification, maintained? If so, do these influence chemical processes such as water/sediment transfers which are perhaps significant in the release or retention of nutrient or toxic materials? It is also important to know the extent to which invertebrate or floral communities are affected, and whether riparian species are influenced by changes in their food supplies.

It should not be surprising that acidic and oligotrophic waters differ from those of circumneutral pH and higher ionic strength. Freshwater ecosystems have been regarded as a 'continuum' (Vannote *et al.*, 1980) with changing physical and chemical conditions from source (low-order streams) to river mouth and of their fauna and flora. Low-order streams are inhabited by a restricted range of species but as distance from the source extends, physical and chemical conditions provide for a greater variety and abundance of organisms (Minshall and Minshall, 1978). Some would claim that it is acidification *per se* that restricts the biological diversity of acid and oligotrophic upland streams, but first-order streams, regardless of pH, have lower diversity than waters downstream. In particular, the variety of habitats (or niches), including the presence of macrophytes, or the absence of sediment substrates, in turbulent upland waters restricts faunal diversity. The size of water bodies also has a significant influence (Fryer, 1985) on zoogeographical distribution along with altitude and climate, accessibility for recolonization, and the availability of food (Willoughby and Mappin, 1988).

Many field studies demonstrate the more restricted fauna of acid waters (e.g. Fryer, 1980; Townsend and Hildrew, 1984; Raddum and Fjellheim, 1984) but these findings are not always explicable in terms of species sensitivity to pH exposure or associated chemistries as demonstrated in laboratory tests. Other factors are also important, including the interrelationships between species in the community, but have been relatively unexplored, other than for predation (e.g. Eriksson *et al.*, 1980), or for application of correlative statistical procedures (e.g. Ormerod and Edwards, 1987). The important interrelationships within the ecosystem, involving physical and chemical conditions, primary biological processes, primary and secondary production, and predation, are simplified schematically in Fig. 10.1

Some field experiments acidifying or mitigating acidity have been reported; these give some insight into the short-term responses of the invertebrate community to acidification but may not reproduce an 'equilibrium' or relatively stable situation that develops over the long term.

10.2   MICROBIAL ACTIVITY—SYNTHESIS AND DEGRADATION

Organic matter degradation, energy flow, mineralization, and nutrient cycling are all critical processes within aquatic ecosystems; microbes, fungi and other micro-organisms have a major role. In some circumstances they also affect the rate of acidification through the generation of alkalinity (Schindler, 1986; Rudd *et al.*, 1986a, b).

One of the early claims of acidification damage to freshwater ecosystems was that bacterial activity was reduced, resulting in progressive 'oligotrophication', with accumulation of detritus and the development of 'fungal mats' (Grahn *et al.*, 1974; Hultberg and Grahn, 1980). However, there are few convincing demonstrations of litter accumulation in lake or stream systems, or of fungal changes (and Grahn's 'fungal mats' were found to be filamentous algae: Hendrey and Vertucci, 1980). However, acidophilic fungi are commoner than acidophilic bacteria and a pH shift from bacterial to fungal decomposers in acid conditions might be expected. Some field studies in acid lakes and streams and field acidification experiments have shown such a shift. In contrast, many other studies report little or no correlation between bacterial plankton and pH, as in the Adirondack lakes (Boylen *et al.*, 1983), Norwegian lakes (Traaen, 1980) and Ontario lakes (Hoeniger, 1985); Traaen suggested that DOC may be more important than pH.

Organic matter is highly variable in freshwater systems, and is the result of three major processes:

— In-wash occurs from streams or lakes upstream, as a result of land use or from erosion; increased upland cultivation, as in southern parts of the English Lake District, has accelerated the transfer of organic matter to lakes there (Pennington, 1984).
— Microbial activity, or slower fungal activity, at the water/sediment interface or within sediments in lakes, breaks down cellulose or other refractory materials; a reduced efficiency is claimed for acidic waters (Traaen, 1980; Rao *et al.*, 1984a).
— Invertebrate species classed trophically as 'shredders' reduce progressively the particle size of biological litter; where the freshwater louse, *Asellus*, is absent, stream litter is slow to break up (Stenson, 1985).

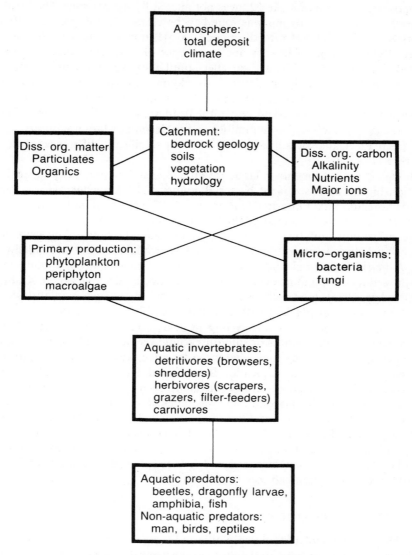

Fig. 10.1—A simplified scheme of ecosystem relationships in aquatic systems.

All these processes might lead to the accumulation of undecomposed litter and lower DOC in acid waters, and the few early observations provided support (e.g. Grahn *et al.*, 1974). However, many inconsistent observations from both field and laboratory studies suggest that it is not universal. Several experimental studies failed to show a change in decomposition rate (Schindler and Turner, 1982; Schindler *et al.*, 1980; Dillon *et al.*, 1984; Brezonik *et al.*, 1986); Schindler (1987) concluded that the rate of decomposition in acidified Lake 223 was not altered by the pH shift. The breakdown of alder leaf litter in Gardsjon was found to be related to the presence of *Asellus* (Andersson, 1985) and not directly related to pH, nor increased by liming. If the reduced decomposition of litter is due to absence of 'shredder' species, it might be expected that liming would have no effect until these species were re-established. However, in Lake 223, most detritus shredding was achieved by acid-tolerant chironomid larvae, and in other sites caddis larvae are sufficiently numerous to break down coarse detritus (Townsend *et al.*, 1983). In some limed lakes, significant and rapid increases in microbial communities and in decomposition are observed, but in limed Loch Fleet and an adjacent acid lake no significant differences were found (Battarbee *et al.*, 1992).

Although deeper sediments in lakes are little changed by the overlying acid water, a relationship between sediment bacterial counts and lake water pH was reported for lakes near Sudbury, Ontario (Hendrey *et al.*, 1976) although not found in a group of acid Norwegian lakes (Traaen, 1978). Experimental studies had suggested a reduced glucose utilization and turnover of phosphate, but attempts to measure a clear relationship between respiration and phosphate availability and pH failed (Laake, 1976). A marked seasonal and year-by-year variation in glucose cycling in Lake Gardsjon sediments found no evidence of reduced bacterial activity there (Gahnstrom and Fleischer, 1985). However, one field study did show such a relationship with respiration reduced at pH 3.8 to 10% of that at pH 7.2 (Francis, 1986).

Reviewing work on major ion cycling in acid waters, Bernard (1990) found no good evidence that phosphorus or sulphur cycling were affected by acidification. The nitrogen cycle appears to be somewhat more sensitive, with inhibition of nitrification below pH 5.5; the rate of denitrification is unaffected. However, Lake 223 showed no net significant change in the pattern or rate of nitrogen cycling (Schindler and Turner, 1982). In lakes where the nitrogen cycle was blocked, nitrification is reactivated with liming after a lag of about a year. On examination of 14 acid lakes in North America and Norway, Rudd *et al.* (1986b) concluded that several decades of acid rain had failed to affect denitrification; however, there is concern that continued deposition of atmospheric nitrogen to catchments could lead to nitrogen 'breakthrough' from the terrestrial ecosystem, and lead to eutrophication.

10.3   PRIMARY PRODUCTIVITY

Species making up the planktonic, epiphytic and periphytic communities are evidently different in acid and circumneutral waters; primary production seems to be unaffected by the presence or absence of particular species, but is primarily related to the levels of nutrients present. As many acid waters are oligotrophic, it is not clear whether acidity *per se* has any effect. In many shallow upland streams, primary production may depend on bryophytes (mosses and liverworts) and epiphytic diatoms; these are more important in fixing carbon than macrophytes or phytoplankton. In lakes, the epiphytic community is still important, but both the littoral or submerged macroalgae and phytoplankton contribute. While the occurrence of particular species and species diversity is related to pH levels, as well as associated chemistry, many species are tolerant of the conditions observed and there is no good or consistent evidence that primary production is affected.

Both survey data on species found in acid lakes, and whole-lake or mesocosm experiments, confirm that species are fewer in more acid lakes, especially at pH 5.6. Some field studies have indicated that dinoflagellates dominate in acid lakes, especially *Peridinium* sp., although in some, Chrysophyta are also often abundant. These acid-tolerant species can be identified in fossil remnants in sediment cores used to reconstruct past lake conditions (see Chapter 7).

Effects of acid conditions on phytoplankton biomass are not at all clear, and reported results are inconsistent; the overall impression is that a lowered biomass is unlikely. Similarly, only negligible changes in primary productivity have been reported. The presence of aluminium or other metals in acid waters may be important, particularly the possible chelation of phosphorus with aluminium, depleting this essential nutrient. The growth rate of *Asterionella* was reduced with aluminium and acid exposure, with H^+ moderating aluminium toxicity possibly through competition for binding sites on cell surfaces (Gensemer, 1991).

The periphyton—microalgae growing on submerged substrates—is an important community in both flowing and lentic waters. In common with other aquatic communities, species richness is reduced in acid conditions, and tolerant species become dominant; some blue–green and diatom species are lost, but green algae increase, especially *Mougeoutia* (Turner *et al.*, 1991). Biomass increases at low pH, but productivity per unit biomass is generally lower.

10.4   ZOOPLANKTON

Zooplankton play an important role in lakes and large rivers in transferring energy from phytoplankton, the primary producers, to the higher trophic consumers, such as fish. The major taxa are cladocerans, copepods and rotifers, groups which are not very important in streams. Zooplankton species are generally fewer and abundance less in lower pH lakes; rotifers are the most sensitive group (Roff and Kwiatkowski, 1977). Limited observations, however, suggest that acidification does not affect plankton grazing rates (Heltcher *et al.*, 1990).

As with phytoplankton the number of species is lower in low pH waters, especially below pH 5.5. In a synoptic survey in the Hebrides, Fryer and Forshaw (1979) found a total of 68 plankton species in 99 low ionic strength waters, of which those at pH $\leqslant$ 5.0 had only nine species. Many other papers (summarized in Heltcher *et al.*, 1990) report similar findings. The same observation is reported for whole-lake acidification; in Lake 223, *Diaptomus sicilis* was lost when lake pH fell to 5.8, and *Epischura lacustris* at pH 5.6, leaving only one calanoid (*Diaptomus*) species. The opossum shrimp, *Mysis relicta*, also disappeared (Schindler *et al.*, 1985). However, a tolerant *Daphnia* sp. appeared, so acidification led to a net loss of only three crustacean species (Malley and Chang, 1985). Elsewhere, in clear acid lakes acid-sensitive species are absent; however, in organic acid lakes there was little relationship of species numbers with pH, possibly due to humic complexes with toxic metals, or to long-term adaptation. The tolerant *Daphnia*, *Diaptomus* and *Bosmina* spp. are the usual zooplankton representatives in acid waters, along with the even more tolerant *Chaoborus* sp., suggesting that these are protected by having a low degree of water permeability (Brett, 1989). Ion regulation and reduced oxygen uptake have been observed in acid-exposed *Daphnia magna*, suggesting that the mechanisms are similar to those of fish (Potts and Fryer, 1979).

Effects such as reduced filtering, gill tissue damage, reduced sodium uptake, haemoglobin loss, reduced oxygen uptake and ecdysis mortality are reported. In the longer term, longevity, brood size and reproductive capacity, and delayed reproduction have been observed. However, a significant relationship with pH could be found in only 14 of 44 studies and the mechanisms are not understood (Brett, 1989). It is suggested that stress, associated with some chemical factors, reduces the relative fitness of a species in acid waters. The effects of acidification on biomass and abundance are only poorly supported in the absence of metal contamination; indeed several studies suggest that acidity has no effect (Heltcher *et al.*, 1990). Acid and aluminium exposure (both in field conditions and in experiments) have adverse effects on some species; Havens found zooplankton species were reduced and that the combination of acid and aluminium was more toxic than either alone at pH 4.5 and 180 μg Al l^{-1} (Havens and Heath, 1989). At pH 5 and 200 μg Al l^{-1}, two daphniid species suffered 100% mortality, while in those conditions the more tolerant *Bosmina longirostris* survived (Havens, 1990).

Biotic interrelationships must play some part in developing an acid community structure, especially if there are changes in predation (*cf.* fish and aquatic beetles, Battarbee *et al.*, 1992), but few studies have been reported.

10.5   MACROINVERTEBRATES

Stream and benthic invertebrates have been studied more intensively than plankton in acid waters. They are an important community for recycling of nutrient materials in the ecosystem, and provide food for many predatory aquatic and riparian species. They are considered useful bio-indicators of acidification or recovery since their response to acid–base chemistry is known for a number of species. A large number and variety of regional or monitoring surveys have been reported.

A consistent picture emerges of the absence of sensitive species as pH falls—these include some mayfly larvae (e.g. *Baetis* spp.), amphipods (e.g. *Gammarus* spp.), crayfish (e.g. *Orconectes* spp.) and molluscs (e.g. snails and clams) (Figs 10.2 and 10.3). These differences are found even at pH 6.0 in low-calcium waters. Experimental acidification of streams and lakes confirm this finding. However, at this moderate level of acidity, tolerant species replace those lost, so that indices of species richness, diversity, or biomass may not reveal a difference from circumneutral waters. In addition, the availability of suitable substrates, the difficulty of finding suitable matching 'control' sites, and geographical and year-by-year seasonal fluctuations make interpretation difficult in respect of populations and communities.

Benthic macroinvertebrates in 20 streams in southern Sweden, with a range of pH and humic content, show a relationship between pH and species richness, with a discontinuity at about pH 5.7, below which the numbers of taxa decline sharply

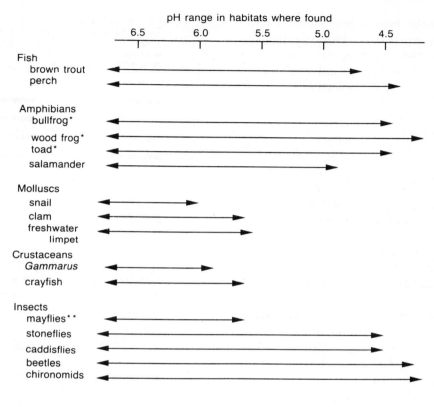

Fig. 10.2—The distribution of various classes of aquatic species through the range of freshwater pH.

(Kullberg, 1992); humic content also has an adverse effect. In acid fishless waters, however, there may be a wealth of other predators, such as beetles, predatory dragonfly larvae and corixids. Corixids and beetles are at an advantage since as air-breathers their sensitivity to water quality is low, and as predators they can profit by the absence of fish. In the limed Loch Fleet, beetle abundance and diversity were reduced (from 36 to < 10 species), and after the reintroduction of trout, selective predation of *Hydroporus palustris* was observed, with *Potamonectes elegans* displacing it in open water (Battarbee *et al.*, 1992).

Stony upland streams are characterized by a diverse community of insect larvae, including mayflies (Ephemeroptera), stoneflies (Plecoptera), caddisflies (Trichoptera), dragonflies (Odonata) and two-winged flies (Diptera). There are also a number of adult insect groups represented, including beetles (Coleoptera) and bugs (corixids, water boatmen and pond skaters). In streams of moderate alkalinity and pH > 6.0, the fauna is rich in total species, especially seasonally. Crustaceans such as *Asellus* and molluscs such as *Lymnaea peregra* and *Ancylus* are also to be found on suitable

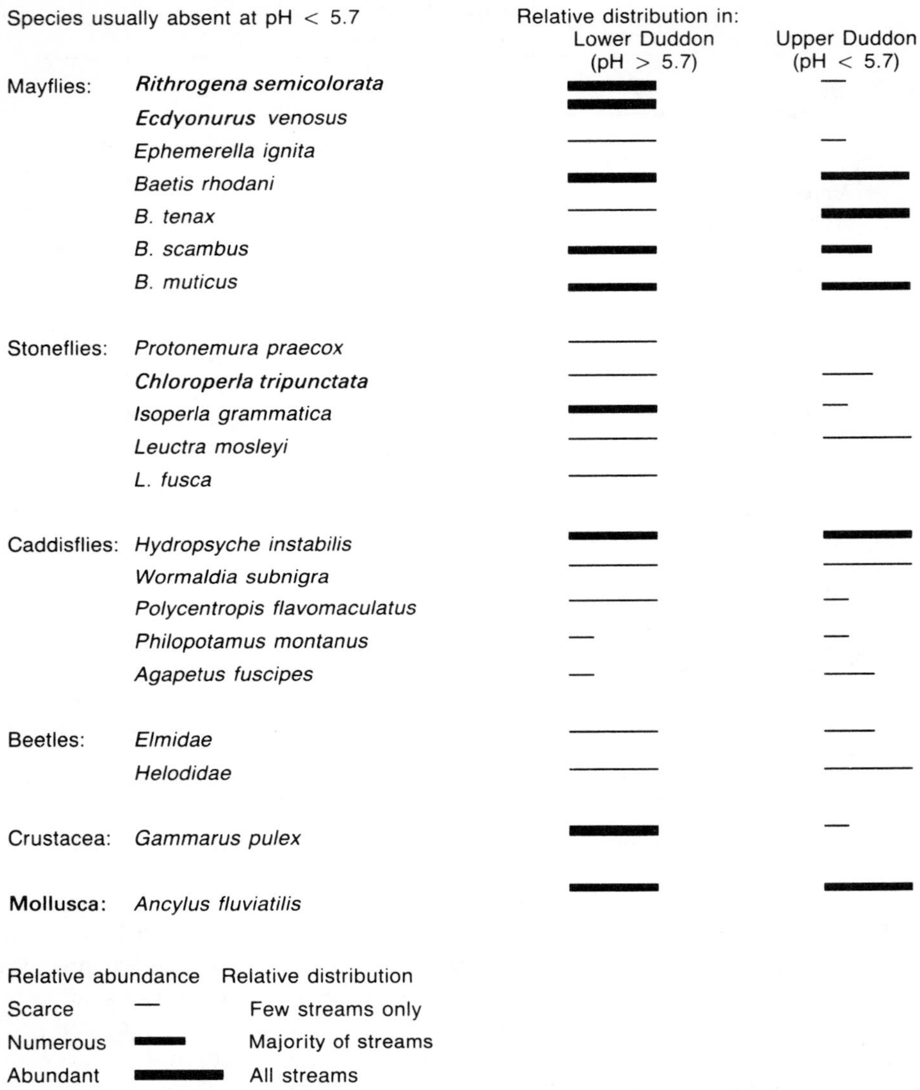

Species usually absent at pH < 5.7		Relative distribution in: Lower Duddon (pH > 5.7)	Upper Duddon (pH < 5.7)
Mayflies:	*Rithrogena semicolorata*	▬▬▬	—
	Ecdyonurus venosus	─────	—
	Ephemerella ignita	─────	
	Baetis rhodani	▬▬▬	▬▬▬
	B. tenax	─────	▬▬▬
	B. scambus	▬▬	▬
	B. muticus	▬▬	▬▬
Stoneflies:	*Protonemura praecox*	─────	
	Chloroperla tripunctata	─────	─────
	Isoperla grammatica	▬▬	—
	Leuctra mosleyi	─────	─────
	L. fusca	─────	
Caddisflies:	*Hydropsyche instabilis*	▬▬	▬▬
	Wormaldia subnigra	─────	─────
	Polycentropis flavomaculatus	─────	—
	Philopotamus montanus	—	—
	Agapetus fuscipes	—	─────
Beetles:	*Elmidae*	─────	─────
	Helodidae	─────	─────
Crustacea:	*Gammarus pulex*	▬▬	—
Mollusca:	*Ancylus fluviatilis*	▬▬	▬▬

Relative abundance		Relative distribution
Scarce	—	Few streams only
Numerous	▬▬	Majority of streams
Abundant	▬▬▬	All streams

Fig. 10.3—The distribution of invertebrate species in streams of the Duddon catchment, UK (after Sutcliffe, 1983).

substrates. In streams of lower pH, the mayflies, crustaceans and molluscs are absent or scarce (Sutcliffe, 1983; Raddum and Fjellheim, 1984), although a tolerant mayfly (*Leptophlebia*), stoneflies and caddisflies remain. For the mayfly *Ephemerella ignita*, absence from upland acid streams is attributed to low alkalinity and the absence of food items since it is tolerant to low pH (Willoughby and Mappin, 1988). Some species of molluscs are clearly absent if calcium is insufficient for shell formation; a survey of a thousand lakes in Norway (Okland, 1992) found no snails at pH < 5.2, but significant factors, including physical conditions, chemistry (particularly calcium) and the presence of predators or parasites influence their occurrence at pH 5.2–6.5. Other impoverished groups may be unable to tolerate the acid and associated conditions, such as the higher level of trace metals in acid waters, or lack of a suitable habitat, e.g. their host macrophytes may be limited by altitude (Stokoe, 1983). For trace metals, invertebrate tolerance is usually greater than that of fish; in acid Ontario lakes, which have high metal concentrations, mercury, lead and cadmium, as well as aluminium residues, are enhanced (Wren and Stephenson, 1991). Although metal concentrations are often greater in the more acid waters, toxicity is not usually greater in acid conditions; indeed for many it is reduced (Wren and Stephenson, 1991) (see also Chapter 8).

10.6   FIELD EXPERIMENTS

Several manipulations of water bodies to stimulate long-term or episodic acidification have been undertaken, and provide insight into community- or ecosystem-scale responses.

Norris Brook, New Hampshire, USA, was acidified from pH 6.0 to pH 4.0 by the addition of sulphuric acid over a four-month period (Hall *et al.*, 1980). This increased the drift of invertebrates over the initial 10 days, although later, little difference was seen in the communities of acidified and adjacent 'control' reaches, although some fall in invertebrate diversity was found. Growth of a mayfly, *Ephemerella funeralis*, was slower, producing fewer larvae after adult emergence in summer, and reducing the potential of the next generation (Fiance, 1978). Addition of aluminium also reduced the pH and increased DOC (Hall *et al.*, 1985) and a copious foam was generated, attributed to the lower surface tension of the treated water, perhaps explaining the enhanced drifting.

A similar experiment in a Welsh stream with added sulphuric acid reduced the pH of a reach downstream from 7.0 to 4.2, and aluminium sulphate added further down the reach raised concentrations about tenfold (Ormerod *et al.*, 1987b). These conditions were maintained over only 24 hours. Three invertebrates tolerated the 300-fold increase in acidity, but mayflies (*Ecdyonurus* and *Baetis*) and *Gammarus* showed 30% mortality. Drift increased, particularly in the lower aluminium-treated reach.

The small oligotrophic Lake 223 in southwestern Ontario was acidified by the addition of sulphuric acid over an eight-year period, reducing pH from 6.8 (1979) to 5.0 (1983), a rate of about 0.25 pH units per year (Schindler and Turner, 1982; Mills and Schindler, 1986). During acidification, lake sediment bacteria reduced much of the added sulphate to sulphide, generating alkalinity which neutralized about a third of the added hydrogen ions. In 1984, the experiment was reversed; the natural regeneration of alkalinity would have returned the lake to circumneutral conditions within two years, but some acid dosing was continued to provide a slower rate of recovery. The sequence of biological changes during acidification (Table 10.1) includes a shift in the phytoplankton community with increase in production and standing crop, the loss of several large benthic crustaceans including the crustacean *Mysis*

Table 10.1—The sequence of changes during artificial acidification and
recovery of Lake 223, Ontario

Year	pH	Changes observed
1976	6.8	Normal community of oligotrophic lake
1977	6.13	Green algae increase, more chironomids, fatter trout
1978	5.93	More chironomids, *Mysis* nearly extinct, copepods lost, minnows fail to recruit
1979	5.64	*Mougeotia* increase, crayfish fail to recruit, minnow almost extinct, 0+ trout increase
1980	5.59	More dinoflagellates and blue-greens, crayfish diseased and fail to recruit, trout fail to recruit
1981	5.02	Primary production > normal, more Cladocera, crayfish fail, sucker and trout fail to recruit
1982	5.09	Productivity normal, zooplankton biomass normal, macrophyte overgrowth, crayfish almost extinct, chironomids decrease, fish fail to recruit
1983	5.13	Productivity/biomass stable, crayfish and mayflies lost, fish fail to recruit
1984	5.4	Dinoflagellates and blue greens still dominate, some zooplankters reappear, algal mats disappear, small cyprinid fish species recover
1987	5.4	Phytoplankton species reappear, zooplankton decrease with fish predation, trout condition good, but poor recruitment
1988	5.8	Many invertebrate species reappear but crayfish, *Mysis*, minnow, sculpins still absent, trout recover but numbers less than before acidification

From Schindler *et al.*, 1985; Schindler, 1989; Schindler *et al.*, 1991; Davies, 1989; Mills and Schindler, 1986.

relicta, a relict species isolated by the retreat of ice-age glaciers, and the decline and loss of crayfish and some fish species. There was increased growth of the filamentous alga, *Mougeoutia*, at about pH 5.6, and the loss of a mayfly, *Hexagenia*, at about pH 5.1. Cladocera came to dominate the zooplankton with replacement of *Daphnia* by *Bosmina*. Recovery from acid conditions has allowed some of the original fauna and flora to return rapidly; re-establishment of fish and crustaceans is slower (Schindler *et al.*, 1991).

Another site, Little Rock Lake, Wisconsin, was separated into two basins in 1985, with one receiving sulphuric acid to reduce pH from 6.1 to 4.7, the other 'control' basin maintaining its pH through to 1990 (Brezonik *et al.*, 1991). Again natural alkalinity generation neutralized about two-thirds of the added acid. Acidification had little effect on N and P cycles, although DOC was reduced, trace metal concentrations increased and sulphate reduction increased with the higher S input. Following acidification (Table 10.2), surface blooms of the blue–green alga, *Anabaena*, previously common, were cleared, with increased diversity of the phytoplankton. Zooplankton biomass did not fall, and species changed little, but daphniid species

Table 10.2—Biological changes during artificial acidification of Little Rock Lake, Wisconsin, by comparison with control basin

Year	pH	Changes observed, relative to reference basin (pH 6.1)
1985/1986	5.6	Increased benthic algae, especially *Mougeotia*, extending to 12–14% of littoral Increased transparency No change in chlorophyll, or ^{14}C uptake Initially no change in phytoplankton species; later decline in chrysophytes Little change in zooplankton: rotifers and crustaceans No change in abundance or species of benthic invertebrate community No change in macrophyte species or distribution Decreased fungal growth, reduced rates of leaf decomposition for one of four species No change in reproductive activity of four fish species
1987	5.2	Further extension of benthic algae Loss of *Leptodora kindtii* No change in abundance of *Chaoborus*
1989/90	4.7	Algal mat covers 27% of littoral Dominant rotifer species changed *Chydorus* and *Oxyethra* (grazing algal mat) increased Largemouth bass fry absent and perch fry reduced sharply

declined. The benthic community invertebrates declined, including mayflies, but only at pH 4.7; invertebrate species here were evidently less sensitive than the community of Lake 223 which showed substantial changes at pH 6.0 to 5.0. Fish in Little Rock Lake (largemouth bass, rock bass and yellow perch) showed reduced recruitment, although perch continued to reproduce at pH 4.7 (Webster *et al.*, 1992).

Amelioration of acid conditions by liming has also brought changes in aquatic flora and fauna in several cases; these are reviewed in Chapter 11. Although acidification and recovery observations in whole lake/stream systems are clearly different, no doubt due to their different characteristics, they do provide models from which response to acidification and amelioration can be predicted, as well as a useful management tool.

10.7 MACROPHYTES

Macrophytes include emergent, floating and submersed forms; they are important as a food source and also provide the conditions for breeding, feeding and nursery areas for fish and some invertebrates. An early report of acidification effects in Swedish waters was the increased abundance of underwater *Sphagnum* moss in lakes of pH < 5.0, where it was thought to have replaced *Lobelia–Isoetes* communities (Grahn *et al.*, 1974). The extent of *Sphagnum* growth in Swedish lakes was related to the degree and timing of acidification. *Sphagnum* is acidogenic, taking up calcium and other ions in exchange for H^+ and through the metabolism of COO^- from organic acids and recycled CO_2 in the absence of HCO_3^- (Wetzel *et al.*, 1985). In *Sphagnum* bogs, cation exchange is calculated to account for about a third or more of the input of acidity (Clymo, 1984); as a consequence the runoff acidity from such bogs is significantly greater than that of rain.

In North America, field surveys have led to conflicting results (Heltcher and Bernard, 1990), with some reporting fewer species in acid lakes, but others finding a response to several pH-related factors (pH, alkalinity, calcium and conductivity) (Jackson and Charles, 1988). In several surveys, alkalinity and calcium seem more important than pH alone. The presence of high organic acid content and possible trace metal contamination confounds the pH relationship.

In Europe, regional survey data are fewer. In Finland species richness in 135 lakes was related to alkalinity (Heitto, 1990). In the English Lake District, species richness was reduced, primarily due to altitude, although soft (low-calcium) water also had fewer species than hard (high-calcium) ones (Stokoe, 1983; Fig. 10.4). Permanently acid waters had 12–20 species at < 200 m altitude, but five or fewer at > 350 m. Lakes with high alkalinity have twice the number of species found in low alkalinity waters, but very large lakes (at lower altitude), even with low alkalinity, have the largest number of species. Although there is general agreement that acid waters have fewer species, inconsistent records of species loss, decline and increase are evident in what appear to be quite similar conditions (Farmer, 1990); evidently, other factors than acid water quality play a significant role.

The European lakes investigated were generally dominated by *Sphagnum* and *Juncus bulbosa*, but this is not characteristic of North American lakes. Experimental acidification of a stream to pH 4 led to increased moss biomass (Hall *et al.*, 1980) and the liverwort *Scapania undulata* is dominant elsewhere in acid streams. Where dense mats of *Sphagnum* occur, it is claimed that some amphibians, fish and invertebrates are affected (Dillon *et al.*, 1984). Liming studies in Europe (Sweden and the UK) have shown that *Sphagnum* growth is reduced as pH, calcium and alkalinity are increased (Raven, 1989) with replacement by other macrophyte species. Thus the *Sphagnum* dominance is reversed. Similar observations come from limed Swedish lakes (Eriksson *et al.*, 1983; Hultberg and Andersson, 1982).

There are few historical records of recent change in species occurrence or biomass; for Galloway lochs, limited records obtained in 1910 and 1964 have been compared with recent survey data (Raven, 1985, 1988). No significant differences were found although there is independent evidence of acidification of the area. Presence of macrophyte spores in sediment cores may indicate changes in abundance of some species, but the confounding effect of land-use changes, such as afforestation or upland farming, makes interpretation uncertain (Farmer, 1990). The firmest evidence of macrophyte response to acidification is from a study in the Netherlands (van Dam *et al.*, 1988).

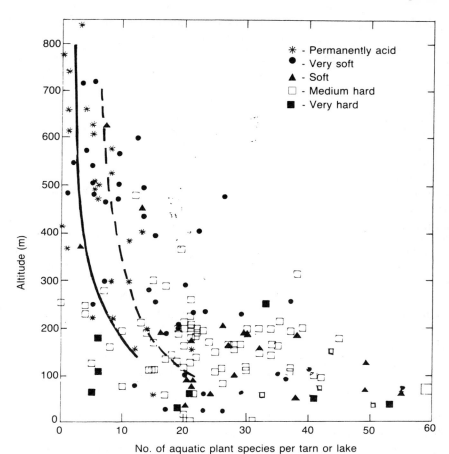

Fig. 10.4—Distribution of macroalgae in lakes and tarns in Cumbria, UK, in relation to altitude and hardness (after Stokoe, 1983). Regressions are drawn from 30 acid (——) and 30 soft water (---) tarns.

10.8   AMPHIBIA

Ampibia may be an important although localized and often seasonal component of the aquatic community. The Anura include frogs and toads, and the Caudata, salamanders and newts. Many temperate species spawn in temporary or permanent water bodies where development from larva to adult occurs. Their exploitation of transient waters for spawning in spring, when highest levels of acidity and aluminium are typical, has stimulated some field and experimental studies of their sensitivity to acid conditions. It is evident that different species have different thresholds of sensitivity and this is reflected in the absence of some species from acid waters. Some species exploit acidic *Sphagnum* bogs or acidic forest soils.

No published studies have been reported on the sources of acidity in temporary ponds; it cannot be assumed that they have been acidified by acid rain, as some are

Table 10.3—Lethal and critical pH levels for amphibian embryos, as determined by laboratory bioassay

Species	Lethal pH	Critical pH	Aluminium LC_{50}
Anura:			
Hyla spp.	3.4–4.2	3.8–4.3	
Bufo americanus	3.8–4.2	4.0–4.2	850 μg at 4.2
Bufo bufo	4.25	4.5	
Bufo calamita	4.5	4.75	
Xenopus laevis	3.0–4.1	3.5–4.5	
Rana arvalis	3.5	4.0	Not toxic
Rana catesbeiana	3.9	4.3	
Rana esculenta	4.0	4.5	
Rana pipens	4.2–4.5	4.6	403 μg at 4.8
Rana sylvatica	3.5–4.0	3.9–4.25	> 200 μg at 5.0
Rana temporaria	4.0–4.25	4.5	< 2700 μg at 4.5
Rana utricularia	3.4	3.8	
Caudata:			
Ambystoma jeffersonianum	4.0–4.5	4.5–5.0	
Ambystoma maculatum	4.0–4.5	4.5–5.0	
Ambystoma texanum	4.0	5.0	

Lethal pH causes 100% mortality, and critical pH significantly increases mortality above controls; aluminium concentrations in μg l^{-1} at the pH where it is most toxic (Freda, 1990).

clearly acidic for other reasons such as *Sphagnum* growth (see above). However, acid snowmelt in spring may lower pH initially, with later rise and alkalinity increase as the year progresses. In some cases, heavy rain has lowered the pH of ponds, but for others there is no correlation between the pH and the volume of rainfall (Freda, 1990). In three areas in North America (Ontario, Nova Scotia and Pennsylvania) 10–15% of temporary ponds have pH < 4.5 and ANC is < 40 μeq l^{-1} (Freda, 1990).

Acid rain was suggested as the cause of high mortality of spotted salamander (*Ambystoma maculatum*) embryos in acidified ponds in New York (Pough, 1976), but this is not confirmed by later studies. Nevertheless, field studies have shown a correlation between pond pH and amphibian abundance, for example in Ontario (Clark, 1986; Freda et al., 1990) and Quebec (Gascon and Planas, 1986); dissolved organic materials, as well as pH, are important. Similar reports come from Europe (Beebee and Griffin, 1977; Leuven et al., 1986) but other authors found no relationship, including the normal development of spotted salamanders in acid conditions (Dale et al., 1985; Cook, 1983).

Most species of anurans are killed in waters of pH 3.5–4.0, but many are able to breed in the range of pH 4.0–5.0 (Freda, 1990; Gosner and Black, 1957) (Table 10.3). Some species can even hatch and survive at pH ⩽ 3.7 (Freda and Dunson, 1984). Experimental studies have shown response to acidity and associated factors; pH, low calcium, high aluminium and DOC are all implicated; such conditions are found at spawning sites in the UK (Cummins, 1986; Tyler-Jones *et al.*, 1989) or simulated in acidification experiments (Clark and Hall, 1985). Both laboratory and field studies have demonstrated embryo sensitivity to aluminium (see Chapter 8), although in very acid conditions aluminium as well as calcium can ameliorate acid toxicity (Freda and Dunson, 1985). As with fish, ion regulation is affected with calcium countering the adverse pH and aluminium effects (Freda and Dunson, 1985; Gascon *et al.*, 1987). Effects are mostly seen during embryo development or when the vitelline membrane fails to expand to accommodate the growing embryo. The hatching enzyme, as in fish, is reported to be deactivated at pH 4.0 (Urch and Hedrick, 1981). During the larval stage, effects are similar to those of fish, namely disruption of the ion regulatory system (see Fig. 8.5 and Chapter 9).

The ecological significance of the failure to hatch amphibian embryos is unclear— the pronounced seasonal nature of the life history means that hazards are high, independent of acidity. Year-by-year differences in timing of snowmelt, heavy rains and temperature changes will occur and predation is often substantial; losses are matched by high rates of reproduction, and poor conditions in one year may be compensated in the following one. Evidently, the wide range of sensitivity of amphibian species allows opportunistic exploitation of a wide range of temporary and permanent habitats, and evidence of large-scale extinction of any species in any geographical region has not been reported. In addition there is strong competition between species within the same habitat; survival is related to species and size (Wilbur, 1984).

10.9 RIPARIAN FAUNA

Riparian birds and waterfowl have been identified as potentially at risk from acidification, either directly or through secondary biological changes. There is also a possible effect (through lack of fish prey) on carnivorous mammals such as otters.

Waterfowl populations in North American lakes have declined in recent years (Scanlon, 1990), probably due to fish loss in acid waters, although data from the field are limited. However, some species may benefit from lack of competition from fish for their prey; many species are omnivorous and can use both terrestrial and aquatic food sources. An experimentally acidified wetland system (from pH 6.8 to pH 5.0) resulted in reduced survival and growth of black duck (*Anas rubripes*) juveniles (Haramis and Chu, 1987), but in contrast, in two acid ponds in Maine, the same species grew faster than in adjacent neutral pH pools (Hunter *et al.*, 1986). As for many reports of acidification effects, it seems that stresses additional to acidity are important. Several authors have reported enhanced metal accumulation in waterfowl living on acid lakes, but the degree to which acid conditions contribute is not clear for such mobile species. Domestic duck species fed on diets with unrealistically high levels of aluminium and calcium have shown interference with phosphate metabolism (with limited phosphate), as well as growth reductions (Carriere *et al.*, 1986). Growth and reproduction of turtle doves (*Streptopelia turtur*) was also affected in more realistic conditions with aluminium accumulating in skeleton and soft tissues. There is also evidence that wild waterfowl abundance is affected by acid water conditions, but other factors such as shore morphology, availability of suitable feeding areas, or lack of fish prey, seem more plausible.

Passerine species consuming aquatic invertebrates have been shown to be affected by the lower species richness of acid streams in Europe. The dipper, *Cinclus cinclus*, is totally dependent on the aquatic environment and adults feed their young on specific size-limited items. In some Welsh upland streams, the territory occupied is extended to provide sufficient food items, so the population density is smaller. There is also evidence that numbers have declined in a group of streams in Wales, in parallel with a decline in alkalinity over the past 30 years (Ormerod, 1985). Blood chemistry reflects limited calcium reserves and body condition is inferior (Tyler and Ormerod, 1992). On the other hand, the grey wagtail, *Montacilla cinerea*, also a riparian species, does not feed on aquatic insect larvae, and their abundance does not show the same relationship (Ormerod and Tyler, 1989). Reduced breeding success and associated eggshell thinning was reported for pied flycatchers, *Ficedula hypoleuca*, in the lake littoral at a non-acidified site in northern Sweden (Nyholm and Myrberg, 1977), leading to speculation that this could be a consequence of aluminium residues in the prey (Nyholm, 1981). However, there is no good evidence that aluminium is accumulated in the invertebrate prey (Wren and Stephenson, 1991) and indeed aluminium concentrations are often lower in biological samples from acid waters, in spite of higher water concentrations. Further, the aluminium found at analysis is likely to be adsorbed on surfaces or present in gut contents and any accumulated during larval development will be shed at moult, leaving none in the adult (Otto and Svenson, 1983).

Concern has also been expressed regarding the effects of acidification on riparian mammals, specifically otters, *Lutra lutra* (Chamier, 1987) and water shrews, *Neomys fodiens* (Warren *et al.*, 1988). The depletion of otters in the British habitat seems more related to loss of habitat than to any other factor. Even in Galloway, Scotland, perhaps the most acidified area in Britain, and where otters are thought to be seriously depleted (Fry and Cooke, 1984), the downstream farmed areas are those most affected, and some otters are still found in the more acid upland areas; this is one of the last areas in Britain where active otter hunting is still pursued (Chanin and Jeffries, 1978). Otter activity was found to be lower near a stream of low pH elsewhere in Scotland, perhaps due to its low productivity (Mason and MacDonald, 1987). In the last few decades, otters are also thought to have been seriously affected by pesticide residues (Warren *et al.*, 1988).

In North America, otters, mink (*Mustela vison*) and raccoons (*Procyon lotor*) also feed on aquatic prey; again it is probable that their abundance will be linked to loss of fish populations in acid waters. A decrease in Norwegian feral mink has been linked with acid rain; however, no data are available to test this hypothesis.

10.10   UNRESOLVED ISSUES

The complex nature and our limited understanding of small oligotrophic waters means that many of their responses to acidification or recovery remain unexplained.

In particular, relatively few species, other than fish, have been studied intensively for their response to acidity and related water quality. There is only a general picture of relative sensitivity of different groups, but closely related taxa have often very different responses.

The responses of communities or of small lake or small stream ecosystems are also inadequately understood; some individual sites have been well studied but generalizations are uncertain. The complex interactions between biological communities and their physical and chemical environments are barely visualized.

Many laboratory experiments do not match field observations, partly because all important factors cannot be mimicked in experiments, but also because strains of target species often vary in their tolerance and in their previous exposure history.

There are many conflicting or discrepant reports on the extent and consequences of acidification; some may reflect different conditions and fauna and flora of different countries or continents, while others may be due to unidentified factors of significance. In some past reports, speculative hypotheses were not always well justified.

10.11 SUMMARY AND CONCLUSIONS

The flora and fauna of acid waters are less diverse than those of circumneutral waters; some species are absent in acid, low-calcium waters, such as Mollusca and some Crustacea.

In contrast, acid conditions may promote abundance of some opportunistic species, such as the alga *Mougeoutia*, the submersed moss *Sphagnum*, and the dipteran larvae of *Chironomus* and *Simulium*.

Although communities are less diverse, basic biological processes are maintained at least to pH 4; slower fungal activity may replace bacterial detritus breakdown, primary productivity is determined by nutrient availability which is unchanged, nitrification and denitrification are unaffected, and secondary production is carried forward by tolerant species of invertebrates, although grazers may be replaced by shredders.

The characteristics of many upland waters are also important determinants of their communities. Altitude, lake and stream size, riparian vegetation, and physical conditions of flow and substrate are all important.

Effects of acidification on riparian species of amphibians, mammals and birds have been proposed but convincing evidence is lacking, with the exception of the dipper, a species dependent on aquatic prey, which requires larger territories on acid streams.

Long-term experiments acidifying waters and then allowing slow recovery have been able to simulate progressive acidification and recovery responses of their biological communities. They have provided insight into the timescale needed for re-establishing species of interest.

11

The reversibility of acidification

11.1 INTRODUCTION

The process of acidification, its causes and effects, have been the substance of preceding chapters. The story is not complete without some consideration of the potential and possible means of recovery. Many of the investigations reported here, and the understanding developed from them, should have value, not only in establishing biological responses and ecosystem effects, and exploring their mechanisms, but also for predicting how such systems will respond to future levels of acid rain, and to develop strategies for reducing or avoiding the unwelcome effects.

Acidification has evidently been a long-term process, covering geological timescales (Pennington, 1981, 1984), although its rate and its effects have increased in the last 150 years. It seems possible that a successful reversal might be equally protracted, even for that part reflecting natural phenomena, which might be difficult or impossible to change. The complexity and interactive nature of ecosystems also implies a dynamic rather than a stable condition, with no certainty that return to circumneutral conditions will recreate some pristine or previous state. Indeed this is almost impossible to specify, given that historical records are few and possibly unreliable, and almost certainly unverifiable. Pristine or unpolluted systems are usually conceived as being of maximum diversity, of both habitat and species, and more desirable than a 'climax' state in which a community is closely matched to existing conditions, often dominated by one or a few successful species. Yet the latter represents a natural sequence of 'evolution' through the Darwinian struggle for existence between unequal species or individuals and the natural response of successful species to become 'fit' for a specific ecological niche. Indeed, a consequence of this struggle is that no two water bodies are quite the same; we tend to put high value not only on sites with high species richness but, paradoxically, also on those where a few opportunistic species provide a unique assemblage.

Given that a desirable 'pristine' state is impossible to define, and given our insufficient understanding of how to manage the community balance, attempts to 'put the clock back' seem doomed to failure. However, if more limited goals can be identified, ecological management is within our reach; we can define the water quality for restoration of a fishery, we can protect a wetland habitat, and we can conserve specific taxa of the aquatic community.

Much of the historical evidence of recent acidification, and sometimes its deficiencies, has been reviewed in earlier chapters; many of the original concepts have had to be revised in the light of new knowledge of sources, pathways and transformations from emissions through to aquatic systems. Doubts of recovery of the chemical or biological status of affected waters arise from observations that little benefit can yet be seen from the substantial ($\sim 40\%$) reduction of S emissions in Europe since their peak in the 1970s; indeed Scandinavian waters are still developing acid symptoms. In the USA, acidification is increasing, although national emissions have changed little over 50 years. Some explanations put forward are that other acidifying emissions (N sources) have grown while S emissions have fallen, that unmeasured stratospheric or biogenic sources have increased, that geological-scale weathering is exhausted, or that long-term accumulation of sulphur in soils continues to be leached to runoff. The continued interest in acidification, coupled with the substantial international research effort of the past few decades, has increased the rate of diagnosis but has not yet found a cure. Certainly, the simple relationships first proposed, matching sulphate concentrations or fluxes in rain and surface waters, and failing to take account of soil influences, as well as the lack of universality of many field observations, leads one to question whether proposed actions to reverse acidification will be effective and timely.

This chapter will first consider, briefly, acidification in a historical context, and then review how field experiments have been able to develop theories and models of the process of acidification, providing insight into how and when various recovery strategies might be effective.

11.2　THE HISTORICAL PERSPECTIVE

Historical observations, although sparse and unreliable for both water chemistry and aquatic communities, as well as the more certain palaeological records of sediment cores, suggest that many surface waters have become progressively acidified over millennia, particularly so where waters drain from areas with unreactive geology and soils poor in weathered mineral materials (Pennington, 1984). Indeed such conditions can develop even in soils overlying chalk formations. Long-term observations of undisturbed plots at Rothamsted, UK, show that the soils have become more acid with time by the accumulation of acid generated by vegetative growth (Johnston *et al.*, 1986); soil pH has fallen from pH 7.1 in the 1890s to pH 4.2 in the 1980s, nearly a 1000-fold increase in H^+ concentration over less than 100 years! Other acid inputs come from accumulated organic material, leaching or sequestration of bases, and of course from increasing acid deposition at this site (Brimblecombe and Pitman, 1980).

Although soil sources of acidity are now known to be substantially greater than annual, or even several years', accumulated acid deposition, the increased use of fossil fuels in Europe since the 1950s is thought to explain the increased acidity reported for surface waters, and the associated biological changes. The 'titration model' of

Henriksen (Chapter 7) was used to explain why the changes appeared so rapidly in chemically poised (low alkalinity) lakes. However, this concept is inadequate without consideration of the role of soils and their contribution of *in situ* acidity and transfer of stored sulphate. This has important implications for the success of strategies to counter or reverse acidification. A return to the supposed levels of emissions of a century or more ago may reduce the acidity of rain to that now found in remote areas, i.e. to about pH 5.0 (see Chapter 1), but it will not replenish the depleted reserves of alkalinity and bases in soils, nor instantly discharge the accumulated sulphur. Moreover, even at pH 5.0, slow soil acidification must continue at sensitive sites.

11.3   CHEMICAL TARGETS OF RECOVERY

Although pH has been termed the 'master variable' it is in fact a shorthand index of acid–base balances (implying either concentrations or fluxes). The relationship between anion and base concentrations,

$$[SO_4]_{in} + [NO_3]_{in} - [bases]_{in} = [H^+]$$

determines the rain pH, and the equivalent fluxes (concentration × rainfall volume) determine the acid loading; lower loading could be achieved by manipulation of any or all of these concentration or volume terms. In surface waters, however, the pH and expected effects are determined by the concentration balance:

$$[SO_4] + [NO_3] + [organic acids] - [bases] = [H^+]$$

These terms are governed by their flues (concentrations × flows) in runoff and influent streams, their uptake within the aquatic system (biota and sediments), concentration due to evaporation, and their discharge downstream. The possibilities for change in water quality thus include manipulation of the soil and runoff concentrations and flows, or of water quality, or of discharge.

A further identified need for biological recovery is to reach certain 'critical ratios', for example that of sufficient calcium (see Chapters 9 and 10); on the basis of a synoptic study of lakes in southern Norway, a ratio of $Ca^{2+}:H^+$ of > 4 was characteristic of lakes with fish (Chester, 1984). Laboratory bioassays for a variety of sensitive and tolerant fish suggest that higher ratios are needed for some species (J. P. Baker *et al.*, 1990). In the case of aluminium toxicity for freshwater fish, the desirable limit for inorganic monomeric aluminium is $< 100 \ \mu g \, l^{-1}$, or even less for sublethal effects on growth; this is not achieved unless soil pH exceeds 5.0 (Bache, 1992) with reasonable buffer capacity. Reduced toxicity within the water body might also conceivably be achieved by a minimum concentration of silicon (Birchall *et al.*, 1989) and a Si:Al ratio > 0.5, or organic materials $(5–10 \ mg \, l^{-1})$ (Howells *et al.*, 1990) (see also Chapter 8). These conditions perhaps suggest options for soil treatment. A further limiting ratio of $Ca^{2+}:S_{load}$ has been suggested on the basis of ratios of lake calcium concentrations and modelled sulphur loading with diatom-inferred pH (50:1 for acidified sites; 70:1 for non-acidified) for some moorland sites in Scotland (Battarbee, 1990).

Similar developments for defining a critical load for forest soils in Finland have led to calculations of the balance between weathering rate and leaching of alkalinity with a ratio of base cations to aluminium, BC : Al, of 1 at steady state (Sverdrup *et al.*, 1992).

These targets provide a variety of theoretical options for improvement of water quality, or for amelioration of adverse effects. Several approaches, both theoretical and practical, have been proposed on national, regional or site bases.

11.4   FIELD EXPERIMENTS: ACIDIFICATION AND RECOVERY

Lake 223 and Lake 302S, Ontario

Two lakes in the Experimental Lakes Area of Ontario have been subject to manipulation by addition of sulphuric acid directly to the lakes, in the case of Lake 302S to a southern basin, while the northern was isolated for comparable nitric and hydrochloric acid treatments. The area has a wet sulphate deposition of about 5 kg ha^{-1} yr^{-1} and the pre-trial pH was about 6.5. Alkalinity was maintained by the supply of catchment materials and bacterial activity in the lake sediments. Other lakes in the area have pH > 5.6 and alkalinity of about 60 μeq l^{-1} and were judged 'sensitive' on the basis of their alkalinity. Lake 223 was first treated in 1976 and further acid additions were made through to 1982, after which a controlled (by lower acid additions) recovery was followed over six years. Lake 302S was acidified in 1982. Results of the acidification and recovery phases of Lake 223 and Lake 302S are summarized by Schindler *et al.* (1980, 1985, 1991).

Acid additions in both lakes were less than initially expected as the lake sediments provided an effective 'sink' for added acid, but acidification at a rate of 0.5 pH units per year was achieved. The pH of Lake 223 fell to 5.1 and was kept at about that level for three years; that of Lake 302S fell to 4.5, also kept at that level for three years. The recovery of Lake 223 was controlled to a yearly increase of 5 μeq H$^+$ l^{-1}, with a decrease in cation alkalinity, possibly coincident with an independent decrease in catchment weathering (Fig. 11.1).

Chemical recovery of Lake 223 was faster than acidification, nonetheless, but the biological response was slower (see Table 10.1). Alkalinity generation within the lakes was always substantial, with microbial sulphate reduction and algal nitrate retention, and some seasonal releases of iron and manganese as well as calcium, magnesium and potassium from the lake sediments. Phosphorus concentrations and cycling were unaffected. There is some indication that the nitrogen cycle was changed, with increased algal uptake and denitrification, and loss of nitrification at pH 5.5, with increase in ammonium.

Biological changes during acidification and recovery in Lake 223 and Little Rock Lake (see below) are summarized in Chapter 10.

Little Rock Lake, Wisconsin

Little Rock Lake, a seepage lake in Wisconsin, was acidified by acid addition in 1985, with pH falling from about 6.1 to 5.6 over two years, although a seasonal fall in pH to 5.5 had previously been shown, due to decomposition and accumulation of

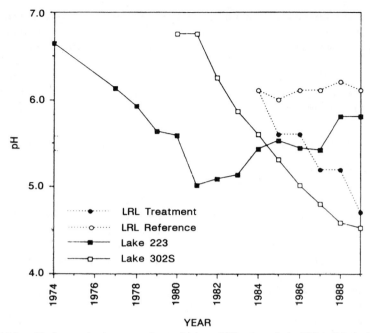

Fig. 11.1—pH changes in three experimental lake acidifications: Lake 223 and Lake 302S, Ontario, and Little Rock Lake (LRL), Wisconsin (after Schindler *et al.*, 1991).

CO_2 below ice cover. Further acid additions brought the pH to 4.7 over a further two years. A summary of the changes is given in Schindler *et al.* (1991) and Brezonik *et al.* (1991); as for Lake 223, internal alkalinity generation was strong due to sulphate reduction and algal nitrogen uptake, and release of cations. Laboratory studies on lake sediments suggest that phosphate was adsorbed to sediments as pH decreased, but there was no inhibition of nitrification at pH 5.2.

Little Rock Lake has a larger number and variety of fish, with an assemblage typical of warmer waters, with dominant largemouth bass and yellow perch; however, as pH fell, survival of juveniles was reduced. The benthic community was relatively tolerant and some species lost at Lake 223 (e.g. *Mysis relicta* and crayfish) were not present at Little Rock Lake before acid treatments. Although the biological responses of this lake are different in detail from those of Lake 223, effects are not inconsistent on the basis of understanding of their somewhat different biological communities.

Sudbury lakes, Ontario

Reducing acid emissions is expected to reverse acidification. The control of emissions at the copper–nickel smelter at Sudbury, Ontario, can be regarded as an experimental test. This site was notorious for its massive exposure to sulphur during the uncontrolled development of metal smelting there which led to severe acidification of soils and lakes nearby, as well as poisoning by heavy metals. In 1972, one major source of sulphur emission was closed and another provided with a 381 m stack to disperse its emission more effectively. Sulphur emissions were reduced from 2.2 Mtonnes yr^{-1} in the period 1950–72 to 0.65 Mtonne yr^{-1} in 1979–83, a decrease of 70%. Air quality was improved and deposition fell by 75%; rain pH rose from pH 2.75 to 3.75 (Hutchinson and Havas, 1986). In two lakes nearby, Baby Lake 1 km southwest of the plant, and Alice Lake 600 m southwest, soil erosion and limited plant growth must have reduced the catchment generation of alkalinity; Baby Lake had reached a pH of 4.0, and both had high levels of dissolved metals. In other lakes in the region, similar but less extreme conditions were seen. Ten years after emissions were reduced, sulphate concentrations in lake waters had decreased and pH increased (Keller *et al.*, 1986; Wright and Hauhs, 1991); in Baby Lake sulphate was halved and pH increased to more than 5.5. Changes in Clearwater Lake, 13 km from Sudbury, compared with changes in Plastic Lake, about 100 km distant, are shown in Fig. 11.2. The rapid response of lakes near to Sudbury might be expected from the high rates of soil weathering and high alkalinity in lake waters there; the more distant Plastic Lake drains a catchment dominated by quartzite.

At Whitepine Lake, 89 km north of Sudbury, similar reductions of acidity (pH 5.4 to 5.9) and sulphate (by 18%) concentrations and rise in alkalinity (1 to 11 μeq l^{-1}) were observed, but also a fall in calcium (by 26%). A return of lost planktonic biota and benthos was observed; species diversity increased but greater fish predation led to a decline in prey abundance. Lake trout recovered although perch suffered from their predation, and residual populations of white sucker and burbot did not recover. Of 104 lakes in the Sudbury area which had lost their lake trout populations, 23 were able to support this species by the late 1980s after restocking (Keller *et al.*, 1992). Although recovery of many components of the aquatic community has followed after 20 years, it has proved impossible to predict the community of the recovering lakes in the Sudbury area. Large numbers of lakes in Ontario remain acidic and contaminated by metals, for which a further reduction in emissions is considered necessary for their improvement (Keller *et al.*, 1992).

RAIN: Sogndal and Risdalsheia, Norway

Some small catchments on sensitive geology in southwestern (Sogndal) and southern (Risdalsheia) Norway have been respectively acidified by enhanced sulphuric and/or nitric acid deposition, or protected from ambient deposition (Wright *et al.*, 1988; Wright and Hauhs, 1991). The RAIN (Reversing Acidification In Norway) experiments continued through a seven-year period, from 1984 to 1990.

At Sogndal, major changes in water quality were recorded (Fig. 11.3). Sulphate concentrations doubled with enhanced sulphuric acid rain, but only by about a third with sulphuric and nitric acid application. About half the increased sulphate was matched by increase in base cations (and alkalinity). Acidity changed rather little— a pH change from 5.5–5.1 ([H$^+$] 3–8 μeq l^{-1}) to 6.0–5.4 ([H$^+$] < 1–4 μeq l^{-1}); alkalinity fell from about 40 μeq l^{-1} to zero. Concentrations of toxic monomeric aluminium increased dramatically with increased sulphate application.

In contrast, the protected small catchment at Risdalsheia showed that complete elimination of the sulphate deposition (or 80% according to Wright and Hauhs, 1991) has led to a substantial reduction in sulphate concentrations in runoff; this fell

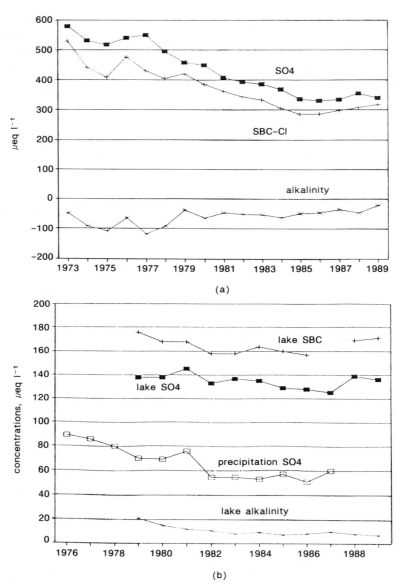

Fig. 11.2—Changes in the chemistry of (a) Clearwater Lake near Sudbury, Ontario and (b)
Plastic Lake, southern Ontario. SBC is the sum of base cations, corrected for chlorides
resulting from road salt icing for Clearwater Lake (after Wright and Hauhs, 1991).

during the earlier half of the period by about 50% compared with an adjacent control
catchment, but since then has maintained a steady output in excess of input, attributed
to slow release of soil reserves of sulphur. Base cation concentrations have fallen by
55% and the critical $Ca^{2+}:H^+$ ratio has fallen; pH and associated water quality

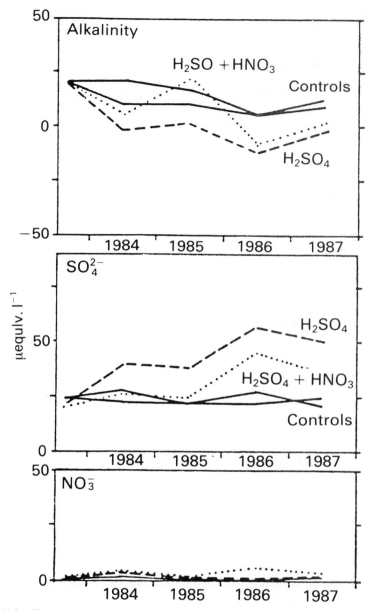

Fig. 11.3—Changes in the runoff water chemistry at RAIN site at Sogndal, with enhanced acid loading (after Wright *et al.*, 1988).

have not sufficiently recovered and the water is still more toxic to salmon and trout juveniles than that from the control catchment. Aluminium fell during the first year, but subsequently was still mobilized from the soil, perhaps because of the increased

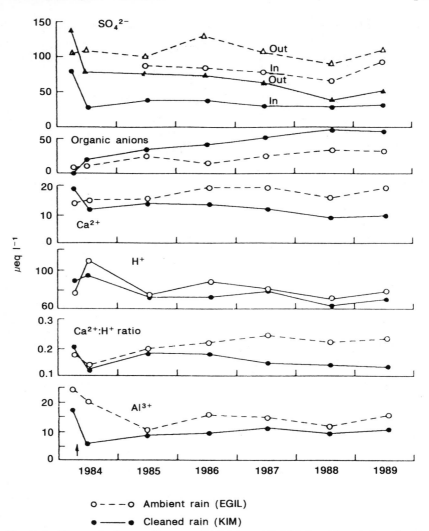

Fig. 11.4—Changes in the runoff water at RAIN site at Risdalsheia, with exclusion of acid loading (data of Wright *et al.*, 1988; after Skeffington and Brown, 1992). The cleaned rain (KIM) is compared with control catchment (EGIL). The graph for sulphate includes inputs.

organic acids present. These chemical changes are summarized in Fig. 11.4 (Skeffington and Brown, 1992).

This unique field manipulation experiment has demonstrated a quite rapid response in these natural systems to both acidification and recovery. Most of the observations are consistent with present understanding of soil responses (see Chapter 5). It is evident, however, that improvement of runoff water quality is insufficient for biological recovery in the short term (almost a decade), even with 80% or more deposition exclusion. With a lesser deposition reduction, an even longer period could be expected;

in addition, soils at the experimental sites are thin, with minimal cation exchange capacity, so a more persistent effect of a soil sulphur reserve might be expected at more typical sites elsewhere.

11.5   CHANGES FOLLOWING EMISSION REDUCTION

Since 1972, several steps have been taken to reduce sulphur emissions in the northern hemisphere, partly due to national commitments to the '30% Club', and subsequently as a result of growing EC and UNECE pressures for control of combustion sources. In the UK, emissions have fallen by 50% since the peak of sulphur emissions in 1965, while for EC countries as a whole, emissions fell by 15% (1978 to 1987); Sweden reduced its emissions by 75% over the same period. Emissions in the USA have been reduced by an estimated < 30% for sulphur and by about 15% for NO_x in the past 20 years (see Chapter 2). It is appropriate here to consider the associated changes to aquatic systems.

Western Sweden
Acidification of the lakes of southwestern Sweden developed progressively through the 1940s to reach a peak in 1970; pH levels were then around 4.5 in many, and sulphate concentrations about 300 μeq l^{-1}; this was accompanied by a reproductive failure in fish populations. To some extent, these changes in water quality reflected earlier hydrological conditions of drought which were relieved in 1977. Coincidentally, wet deposition of sulphate decreased marginally, in spite of a twofold decrease in national emissions (Sanden et al., 1987), but not the acidity of rain, which increased, possibly due to a higher (34%) nitrate deposition (Forsberg et al., 1985). During chemical recovery of the lakes, pH rose by 0.3–0.4 units, with sulphate reducing by 40–100 μeq l^{-1} and calcium and magnesium concentrations decreased (Forsburg et al., 1985; Fig. 11.5). Changes in water chemistry of some lakes were rapid, contemporary with changes in rain composition, in spite of an expectation of a delay for slow soil response. In contrast, a well-documented record over the past 20 years for several Swedish rivers shows no significant change in acidity (Ahl, 1986), although the 0.25 pH unit per year decline from 1965 to 1969 had been predicted to continue.

Since so many lakes in southern Sweden have been limed to reduce their acidity, it is now difficult to judge regional responses there to reduced emissions.

Southern Norway
A slight change in sulphate concentrations of lake and river waters of southern Norway has been reported since 1974 (Henriksen et al., 1988), but has been offset by an increase in nitrate, with little change in pH. The overall change in acid anions is negligible, suggesting that acid soil conditions remain; 50% of lakes still have water quality unsuitable for fish (Skeffington and Brown, 1992), and fishery status continues to decline (Henriksen et al., 1989; Muniz and Walloe, 1990).

At three long-term stream monitoring sites in southern Norway, decreases in sulphate concentration have been 1.7–2.2 μeq yr^{-1}, in calcium about 1.3 μeq yr^{-1}, and in acidity 0.6 μeq yr^{-1} at one site only, with aluminium unchanged. At one of

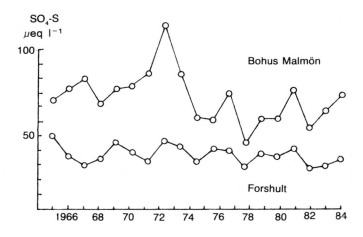

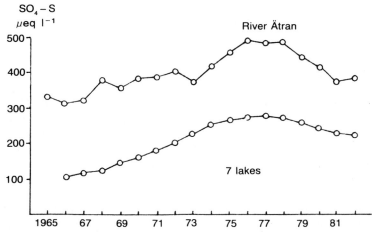

Fig. 11.5—Changes in some Swedish waters following reduced deposition (after Forsberg
et al., 1985).

the sites, sulphate deposition fell by about 30% in the period 1973–87 (Skeffington
and Brown, 1992).

Nitrate concentrations in southern Norwegian lakes appear to have almost doubled
in the period 1979–86, even though nitrate deposition has scarcely changed (Henriksen
et al., 1988). It is suggested that increasing leaching of nitrogen from catchments is
responsible (Wright and Hauhs, 1991).

Scotland

In spite of the substantial decrease in emissions in the UK over the past 30 years,
there is little documented change in quality of sensitive waters. In a comparison of
chemistry of 52 Galloway lakes sampled in 1979 and in 1988, excess sulphate, base

cations and inorganic aluminium were lower at the later sampling date (Wright and Hauhs, 1991). Analysis of 12 sequential samples from two upland lakes over the same period (Battarbee *et al.*, 1988) also suggested first signs of a recovery, consistent with a change in diatom flora. However, the 1979 samplings followed a period of drought, then by a decade of more than average rainfall; the latter period was also characterized by a change in the westerly weather pattern. Some anomalous observations, such as increased calcium with decreased sulphate, are unexplained. At two sites nearby, Loch Dee and Loch Fleet, intensively sampled data for rain and water quality show no significant changes in rain pH or sulphate deposition over the period from 1981 to the present, nor significant change in sulphate concentration of lake waters (Lees and Darley, 1993; Howells *et al.*, 1991). Statistical analysis of data for eight Scottish sites found only one (Eskdaleuir, about 70 km east of Galloway) had a significant ($p < 0.01$) decrease in sulphate, while the mean for all sites was non-significant (Irwin *et al.*, 1990).

North America
In general, a reasonable correspondence between sulphate deposition and surface water sulphate concentrations has been found for both Canada and the USA (see Fig. 6.7; Irving, 1991); in general wet deposition of about $15 \text{ kg SO}_4^{2-} \text{ ha}^{-1} \text{ yr}^{-1}$ leads to about $100 \text{ } \mu\text{eq SO}_4^{2-} \text{ l}^{-1}$ in surface waters. However, no simple or universal relationship is found with acidity and ANC because of the intervention of soils, but an estimated increase of ANC up to $30 \text{ } \mu\text{eq l}^{-1}$ in the Adirondack lakes might result from a 50% sulphur emission reduction (Fig. 11.6). A less well-documented relationship is found for N emissions, nitrate in rain and surface waters since most N species are taken up by the aquatic biota.

At the Hubbard Brook Experimental Forest in New Hampshire, USA, a long record of rain and stream chemistry is available (e.g. Driscoll *et al.*, 1989). Sulphate deposition has declined there by 33% over 22 years, along with a 50% fall in base cations; streamwater concentrations of both sulphate and base cations have fallen by 17%, although the alkalinity deficit has improved. No significant change is seen in stream acidity (Wright and Hauhs, 1991); this lack of response to lower deposition load is interpreted as the result of continuing soil acidity, since nitrate deposition shows no overall sustained increase at this site.

At White Oak Run, a stream in Virginia, subject to a slightly increased deposition load over eight years (1980–87), sulphate concentration increased by 22%, base cations by 6%, and alkalinity halved (Wright and Hauhs, 1991), again indicating continued acidification. Here sulphate deposition is about three times greater than that discharged in runoff; it is thought that the balance is adsorbed on to catchment soils (Cosby *et al.*, 1986) which may now be reaching saturation (Wright and Hauhs, 1991).

In Nova Scotia and Newfoundland, a 25% decrease in sulphur emissions in eastern North America between 1971 and 1982 was matched to a 45% decrease in sulphate flux in rivers (Thompson, 1987), a finding somewhat at variance with studies elsewhere.

Overall, the prognosis from such studies, as well as from modelling and field records, is a disappointing outcome for recovery of fisheries from substantial sulphur emission reductions except for the grossly impacted Sudbury lakes.

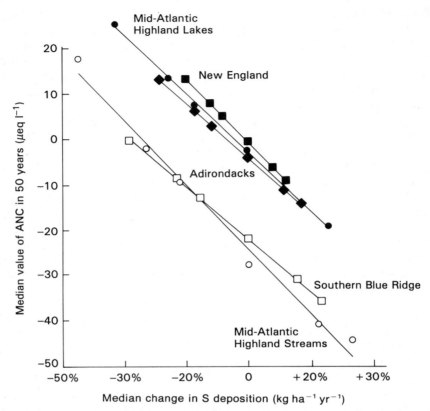

Fig. 11.6—MAGIC-modelled changes in ANC for acid waters in NAPAP regions, with changes in S deposition (after Irving, 1991).

11.6   MODELS OF ACIDIFICATION AND RECOVERY

A variety of models has been developed to reconstruct and predict the course and rate of acidification, particularly for soils and surface waters. The ILWAS and MAGIC models aim to integrate processes and transformations on an ecosystem scale, while others, such as the Henriksen approach (see also Chapter 7), are based on empirical or statistical relationships arising from observations.

Ion balance or titration (Henriksen) models

The concept of the titration model is the analytical tool for estimating acid-neutralizing capacity, the titration of a strong base solution by a strong acid. It conceives of a lake or catchment containing a fixed alkalinity or ANC which is progressively consumed by the input of acidity in rain, analogous to the titration of a base solution against acid in a laboratory beaker. Acidification of sensitive sites (i.e. poor in buffers) is defined as a change from the 'pristine' or historical level of ANC to the current level of ANC. For soils or surface waters, ANC is defined (Reuss and Johnson, 1986) as:

$$[Ca^{2+} + Mg^{2+} + Na^+ + K^+ + NH_4^+] - [SO_4^{2-} + NO_3^- + Cl^- + F^-]$$
$$= [HCO_3^- + A^-] - [H^+ + Al^+]$$

This can be simplified, on the assumptions that only calcium and magnesium are the significant base cations, that sulphate is the major acid anion, and that organic acids are unimportant (Henriksen, 1982):

$$ANC = [Ca^{2+} + Mg^{2+}] - [SO_4^{2-}] = [HCO_3^-] - [H^+] - [Al^{3+}]$$

Henriksen (1980) also found an empirical relationship for lakes in southern Norway:

$$Alk = -14 + 0.93[Ca^{2+} + Mg^{2+}]$$

where base cations are those derived from the catchment, i.e. non-marine sources.

Such simple relationships, however, do little more than demonstrate the validity of the charge balance, and are not necessarily representative of waters other than for the region from which the data are derived. They do not provide an independent approach for testing whether a change in sulphate input to a lake or catchment would change the ANC status or the level of basic cations in the catchment soil or water body. Critical limitations of the model as formulated are that:

— acidification is not directly proportional to the increase in $[SO_4^{2-}]$ in surface water;
— site characteristics and seasonal changes are not included;
— acidification, like the equation, would be reversible;
— cation supply (weathering) does not match the progress (or regress) of acidification;
— many other water quality components are excluded.

There is no place in this model for processes now acknowledged to be important— the role of nitrate or ammonia, retention and/or release of S and N species in soils or lake sediments, soil exchange mechanisms, and production and release of organic acids. In a further development of the model, Henriksen (1982) does consider changes in base cation supply, but argues that they are not influenced by changes in sulphate input. A factor 'F', representing the extent to which a change in sulphate flux is related to a change in base cations, was found to be a maximum 0.4 for both Norwegian lakes and those near Sudbury, Ontario, although the rapid chemical recovery of Sudbury lakes indicates a larger weathering capacity there (Skeffington and Brown, 1992).

This empirical approach has been applied to the 1000-lake database for Norwegian lakes (Henriksen et al., 1988), predicting that a 30% reduction in S deposition would result in a comparable reduction in sulphate concentrations in lakes, with a *pro rata* reduction in $[H^+]$. Henriksen calculates that this would improve alkalinity in 15% of the 700 lakes that at present lack a bicarbonate buffer system. A 50% reduction would improve the lakes so that more than half would have a lower strong acid content and a measurable alkalinity. This prediction was not confirmed by the resampling of 305 lakes previously sampled in 1974–75, a period over which European emissions declined by about 20%. Indeed there has been little or no change in acidity in southern Norwegian waters, although some have shown a small reduction in sulphate concentration, and almost all have experienced a rise in nitrate.

This lack of response suggests that the reality is more complex than the model, perhaps indicating that the reduced acid (or sulphate) input has reduced the ANC supply, and that other chemical constituents have to be included, such as organic acids, or that there is a delayed response. The RAIN experiment suggests that all these effects do occur. Biological response is also disappointing, with continued reports of episodic fish kills (Rosseland et al., 1986) and further loss of fish populations from lakes (Henriksen et al., 1989), although the model suggested that a 30% reduction would restore 28% of lakes and a 50% reduction, 40% of lakes (Fig. 11.7).

An index of the rate and extent of acidification of Scottish lakes has been proposed by Battarbee (1990) on the basis of the relationship of total S deposition and lake-water calcium concentrations. However, the significant input of marine sulphate and base cations at these westerly sites is ignored, and the calculated total deposition requires confirmation by measurements of acid loading which are often inconsistent with modelled values (Irwin et al., 1990). At Loch Fleet, for instance, annual influx in bulk deposition is only 0.52 g S m^{-2} and lake $[Ca^{2+}]$ (pre-liming) was $48 \ \mu\text{eq l}^{-1}$, giving a ratio of $Ca:S_{in}$ of 92 (characteristic of non-acidified sites!) while the nearby Loch Dee has a ratio of 42, characteristic of acidified ones. Even if bulk deposition values are increased by 30% for dry deposit (consistent with in/out flux ratios corrected for evapotranspiration loss), the ratios are 69 and 27. The former loch had a pre-liming pH of 4.5 and no fish, the latter maintains a pH of about 5.6 and a sustained trout population.

The concept that a ratio of $Ca^{2+}:H^+$ necessary for a sustained fishery proposed by Chester for Norwegian lakes (1984) has been developed to model the expected fish communities in North America (J. P. Baker et al., 1990). In addition to pH and calcium, this takes account of aluminium sensitivities of some species coupled with MAGIC model simulations (see below) of water chemistry and baseline factors such as altitude, lake size, and other chemistry. A 50% reduction in deposition is calculated to improve brook trout abundance in Adirondack lakes by 5% and the number of lakes with this acid-tolerant species by 6.5% by the year 2034. For the more sensitive common shiner, the same deposition reduction would have no significant effect.

The current UNECE initiative in developing 'Critical Loads' (CL) for protection of sensitive areas can be seen as a further empirical, target-oriented, approach (Bull, 1991; Hornung, 1993) and is relevant to the recovery of acidified systems. The critical or threshold criterion favoured for forest soils is a $Ca:Al$ ratio, or a $BC:Al$ ratio, of 1 with tree roots the target, while for fresh waters it is a zero alkalinity level or the ratio $Ca:S_{load}$ with brown trout as the target. Alternatively, CL for fresh waters has been expressed in terms of the leaching of base cations from the catchment to provide a minimum ANC ($20-50 \ \mu\text{eq l}^{-1}$) concentration for identified target species (Henriksen et al., 1992). On this basis, mapping on a grid scale for Nordic countries ($150 \text{ km} \times 150 \text{ km}$) and the UK ($10 \text{ km} \times 10 \text{ km}$) has identified areas of 'exceedance'. The Critical Loads approach has still to be refined, and might prove a useful tool. However, it has to be said that there are many shortcomings; selection of a 'most sensitive' target within the unit area, heterogeneity of soils and ecosystems within the unit, uncertainty of how to incorporate N species, and attention to short-term and seasonal effects all need further consideration.

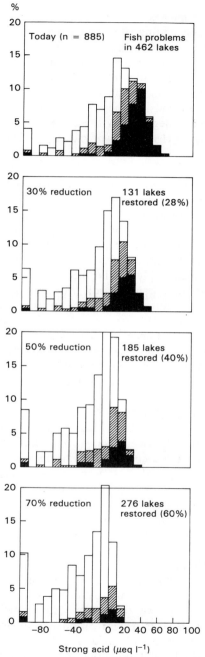

Fig. 11.7—Prediction of fishery status in southern Norwegian lakes with various reductions in lake sulphate concentrations, as a surrogate for reduced deposition (after Henriksen *et al.*, 1988).

Soil-oriented charge balance models

These models include understanding of soil processes and equilibria (Reuss, 1977; Reuss *et al.*, 1987) and have been developed to understand the basic relationship between acid deposition and soil mechanisms, rather than to simulate realistic conditions. They include sulphate adsorption, changes in soil base saturation, and the relationship of ANC and CO_2 equilibrium at different levels of soil acidity. At the high CO_2 pressures within soils, an increase in strong acid anions can switch soil solution ANC from positive to negative. With equilibration to atmospheric pressure, leachate pH can change by a pH unit of 1 or more, possibly explaining some short-term episodes in stream pH.

Integration of dynamic soil and hydrological processes is a characteristic of recent models, including the Birkenes model (Christopherson *et al.*, 1982, 1990; Christopherson and Wright, 1981), DAM (Duddon Acidification Model: Tipping, 1989), and the integrated deposition catchment models, e.g. ILWAS (Integrated Lake-Watershed Acidification Study: Chen *et al.*, 1983; Goldstein *et al.*, 1984) and MAGIC (Modelling Acidification in Groundwater and Catchments: Cosby *et al.*, 1985). Further dynamic models are reviewed by Warfvinge *et al.* (1992). While they have significant differences, they all apply similar assumptions about key soil processes and, implicitly, the mobile anion concept, soil processes including cation exchange, carbonate equilibria, mobilization of aluminium species, and a functional relation between aluminium species and H^+ concentration. They also require information about acid loading, soil and hydrological characteristics, and runoff (output) chemistry, as well as relevant equilibria. Their development and application is critically dependent on sufficient detailed field data, on laboratory determinations of dissociation and equilibrium constants, and on calibration and independent validation in the field of predicted conditions.

The Birkenes model began as a hydrological model to explain episodic variations in Birkenes streamwater quality which was highly dependent on flow; high flow led to low pH and low calcium and vice versa, with aluminium being released at low pH. Later developments included chemical equilibria, although aluminium is only considered as the single species Al^{3+}. It is based on a 20-year record of stream chemistry, with chloride and sodium assumed to be derived from sea salt and to be charge balanced. In spite of its chemical limitations, the model has been useful in explaining changes in stream water over the short term, although an adjacent site exhibits different behaviour, explained in terms of a greater weathering rate. It appears more suitable for short-term responses, rather than to emission controls.

Weathering rate models and ANC mass balance models, including the 'trickle down' model (Schnoor and Stumm, 1986), are based on lake and catchment ANC, including ANC generated by weathering. In this respect, ANC is conceived only in chemical terms and kinetically controlled by input acidity, with net chemical weathering matched to the measured base cation output. In turn, this implies that the forcing variable is the incoming rain acidity and that soils play a relatively small role. It is also assumed that sulphate input is matched by output and thus in equilibrium with any soil sulphur component, which is inconsistent with field experimental observations. Further, while the present exchangeable soil reserve in poorly buffered soils is equivalent to 50 to 200 years of acid deposition at present rates, this ignores the much larger within-soil acid generation (Bache, 1992). The different emphasis of this model, with a lesser role for soil processes, leads to different predictions, although where soils are poor and thin, the rather linear response of streamwater chemistry to changes in input support the model.

The development and application of DAM, based on long-term data for the upper Duddon catchment, simulates conditions there, although it appears to be more sensitive to changes in NH_4^+ inputs than to SO_4^{2-}. Ammonium deposition in this area is high (Irwin et al., 1990) and reductions in sulphate deposition seem likely to be less effective than reductions in ammonium. The 25% reduction in emissions over the past decade led, however, to a 30% reduction in stream sulphate concentrations, which was predicted to increase stream pH by 0.5 units; in fact such an increase has not been reported (Diamond et al., 1992), although reduced agricultural liming in the catchment, and sampling during high but not low flows, may have confounded the possible response.

The ILWAS model is based on an integration of pathways, transformations and processes from deposition through to surface water. Data were derived from three forested catchments in the acidified Adirondack region in upper New York State. Sub-models were developed for deposition, canopy interactions, hydrological and soil and in-lake processes. The model includes deep as well as shallow soil compartments, soil cation exchange with aluminium dissolution and sulphate adsorption, ion balances and carbonate equilibria, as well as biological activity in canopies, soils and sediments. Unfortunately this has made it excessively complex, with problems of validation with independent and sufficiently detailed data sets. A 50% sulphur emission reduction predicted that the pH of the most sensitive site (Woods Lake) would be raised over a decade except during high flow conditions when little benefit would be seen (Gherini et al., 1985; Fig. 11.8), a pattern of stream response characteristic of streams in unacidified regions (see Chapter 7). It is suitable for prediction on an intermediate timescale.

The MAGIC model is a mechanism- and process-oriented model relating deposition (or more strictly national emissions) to surface water chemistry. Soil processes are included quantitatively, such as ion balance equilibria, anion (sulphate) retention, mineral weathering and cation exchange, and CO_2 degassing; these processes are linked with hydrological pathways and flows. The initial development was based on an intensively studied catchment in Virginia, USA, but it has been widely applied to catchments in Scandinavia and Britain and is consistent with current and past (diatom) field chemistry observations (Fig. 11.9; Whitehead, 1989; Wright and Hauhs, 1991). The model is suitable for longer-term predictions (> 100 years) of the response of soil properties and water quality to changes in acid input and land use. Some critical deficiencies are that national or regional S emissions need not represent S or acid deposition at the selected site, that the 'lumped' parameters for soils do not represent the heterogeneity of real soils, and that hydrological parameters vary between sites and seasons. A further problem is that such long-term predictions are difficult to verify in the short term. Nonetheless, it has the merit that data requirements are moderate and that it can be applied to a wide variety of sites.

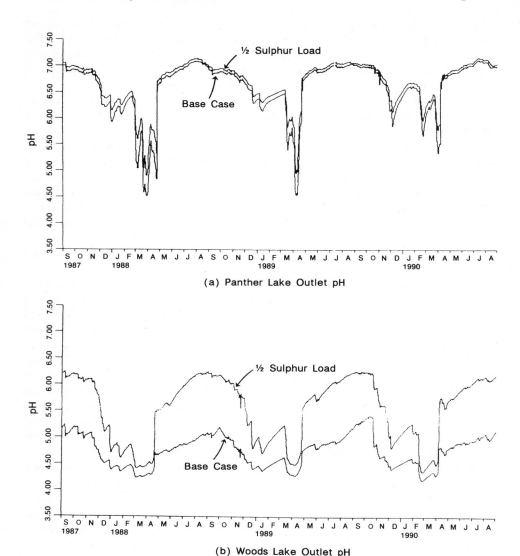

Fig. 11.8—ILWAS-modelled pH in two Adirondack lakes with a 50% S deposition reduction scenario (after Goldstein *et al.*, 1984).

The output of MAGIC provides water quality conditions for a variety of scenarios. Examples for UK waters using a variety of scenarios for emissions and land use are illustrated in Fig. 11.10 (Whitehead, 1989; Skeffington and Brown, 1992). It predicts that a 'no action' scenario leads to continuing acidification, while a 50–60% reduction leads to an initial pH rise, after which it falls slowly. However, if emission control is coupled with liming at a moderate rate, there is continued water quality improvement. A notable feature is the slow response of the system: 150 years for a modest 0.5 pH

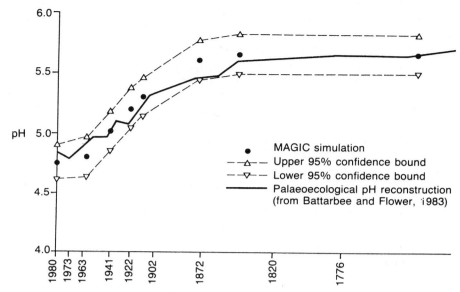

Fig. 11.9—Validation of MAGIC predictions for Round Loch of Glenhead, Galloway, from diatom-inferred pH (after Whitehead *et al.*, 1989).

unit rise even for 80% reduction of national emissions. While aluminium reflects pH changes, it does appear to respond more rapidly, but the calcium response to lower soil acidity means that the critical $Ca^{2+}:H^+$ ratio would make conditions worse for fish and other aquatic biota. According to the model, using Skeffington and Brown's parameters (1992), it would take 245 years for this ratio to rise to 4 with an 80% emission reduction, while catchment lime application will bring it to a satisfactory value almost immediately.

The effects of moorland or forest land use, or the effects of drought (increased evapotranspiration) or increased sea salt scavenging, are also shown (Whitehead, 1989). Again, at current emission levels acidification will continue, but the additional evapotranspiration and sea salt scavenging of a forest canopy would bring a greater pH decline.

MAGIC has also been applied to prediction of changes in the aquatic fauna (Ormerod *et al.*, 1988), with decline of trout populations in sensitive waters in Wales halted with a 50% emission reduction from 1984 levels, given a moorland economy, but no recovery of trout density could be predicted. For the invertebrate communities, the outcome is less certain—in sample data sets for Welsh streams, no evidence of change could be found in the period 1940–84, in spite of evidence of acidification during that period. Moreover, simulated forest development is predicted to impoverish these communities even with reduced emission scenarios.

11.7 ALTERNATIVE STRATEGIES FOR RECOVERY

A number of alternative strategies for soil and lake recovery can be identified (see section 11.3, above), and in some cases have been put into practice.

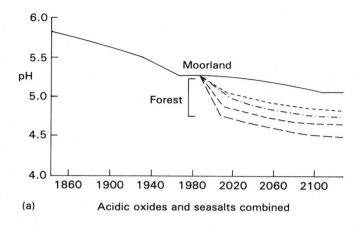

(a)　　　　Acidic oxides and seasalts combined

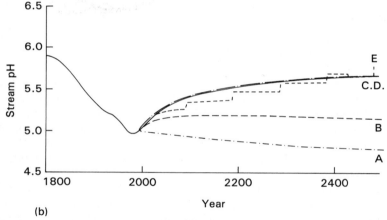

(b)

Fig. 11.10—MAGIC simulations of changes in runoff following changes in land use and management, and emission reduction (after Whitehead *et al.*, 1989, and Skeffington and Brown, 1992); (a) shows the effects on acid loading by enhanced scavenging by forest canopy; (b) shows the effects of various UK emission reduction scenarios—**A** = no change, **B** = 60% reduction, **C**, **D** = 80% reduction, **E** = as for **C**, with 3 t ha^{-1} lime every 100 yr.

Liming

Liming to counter soil acidity has been practised at least since Roman times; it has been used commonly in upland areas as part of agricultural 'improvement' in Britain. At present, 7.4 Mha receives 4 Mtonnes of $CaCO_3$ annually (Bache, 1992), equivalent to 0.5 t ha^{-1}. This is needed to neutralize not acid from deposition but the much larger component of soil acidity resulting from base uptake and biological activity within the soil. This represents a 30% fall from applications made prior to 1976, when liming was curtailed by withdrawal of a national subsidy (Ormerod and Edwards, 1985). Uncultivated uplands have seldom been limed, but are subject to progressive acid accumulation from both deposition and vegetative growth, and a strong case can be made for liming in sensitive areas in Britain and Scandinavia. A few examples are summarized here.

Direct liming of acid waters also has a long history, even before its benefits were understood scientifically. It was also practised early this century to counter effects of mining and industrial discharges. In Sweden, where particularly strong concern was expressed about regional lake and stream acidification, much effort and resources have been given to developing methods and setting up practical schemes for restoring fisheries. More than 5000 lakes in Sweden have been treated, through government funding amounting to nearly 100 million SKr; in Norway, about 200 lakes and streams have been treated also with government support. In North America, a number of trials have been undertaken, especially in the Adirondack lakes and in streams elsewhere receiving acid mine waste or drainage. Research support in the USA has come from government agencies and several reviews and handbooks have been published to guide potential users (e.g. Olem, 1990; Brocksen *et al.*, 1992). The principal objective has usually been to restore or improve fisheries. The use of limestone materials in drums, silo-style dosers and streamflow-operated wells has also been tested extensively (see Olem, 1990).

Many specific examples demonstrate that both water chemistry and fish communities benefit from liming (Eriksson and Tengelin, 1987; Howells and Dalziel, 1992; Howells *et al.*, 1993; Hultberg and Nystrom, 1988; Nyberg, 1984). In Sweden, substantial changes in Gardsjon followed direct lime applications; pH rose from about 4.7 to > 7.0 and associated chemical changes included an increase in dissolved organic carbon and a decrease in aluminium, possibly linked with the decay of *Sphagnum*. A comparison of limed and unlimed streams showed that density of salmonids was increased, but for other species the picture is less clear (Degerman and Appelberg, 1992), probably due to biotic interactions since minnow and burbot increased in some waters but not in those with salmonids. Liming of salmon spawning beds in Nova Scotia streams increased spawning, improved fry survival and increased juvenile density; these benefits were also seen for brook trout (Lacroix, 1992). There is less information about effects on other aquatic biota; in general, acid-sensitive species have reappeared, including sensitive mayfly species (Fjellheim and Raddum, 1992; Battarbee *et al.*, 1992), and dominance of acidophilic species such as underwater *Sphagnum* has been reduced. Liming of shallow temporary moorland pools used by *Rana arvalis* for beeding improved survival, but apparently due to reduced fungal infection (Bellemakers and van Dam, 1992). It has been argued that direct liming is expensive, since its benefits are transient unless lime applications are repeated, and that a stable water quality is not achieved.

An alternative strategy to direct liming is to provide lime to catchments draining to acid waters. This has evident benefit for rapidly draining systems characteristic of uplands; effects are longer sustained, and water quality changes are relatively slight and rather stable, avoiding high conductivity, high calcium, conditions which are inimical to the indigenous fauna and flora.

In Norway, the small catchment of a high altitude fishless lake, Tjonnstrand, was limed in 1983. Lake retention was only three months, so lake liming would not be effective. An application of 75 tonnes (3 t ha^{-1}) brought a rise in pH from 4.5 to 7.1, with pH > 5.8 maintained through the following four years, although heavy rain events still resulted in acid pulses. Calcium levels were raised to 4 mg l^{-1}, and aluminium (total) to about 100 μg l^{-1} (Rosseland and Hindar, 1988).

At Loch Fleet, southwest Scotland, lime applications to the 110 ha catchment were made to selected areas in 1986, including below a forest canopy, surface liming of moorland, and wetland source liming. The lake has a retention time of about six months (Howells and Dalziel, 1992). A total of about 450 tonnes of CaCO$_3$ was applied, overall about 3 t ha^{-1}. Water quality was quickly improved with pH rising to about 6–7, calcium to 4 mg l^{-1}, and labile aluminium greatly reduced (Fig. 11.11). Predictions from calcium budgets indicate that conditions suitable for trout may be maintained over 15 years at this site which has a mean annual acid deposition of about 0.43 g m^{-2} (1985–90) (Dalziel *et al.*, 1992a). Trout restocked to the lake 18 months after the first liming have shown good growth and fecundity (Fig. 11.12); invertebrate communities showed some changes, with decline in predatory beetles, undoubtedly due to fish predation, and return of some acid-sensitive species.

A similar liming application made to upland areas in southwest Wales included two moorland sites, one over the whole catchment (9 t ha^{-1}), the other in a selected source area only (30 t ha^{-1}). The treatment resulted in improvement of streamwater quality (Donald and Gee, 1992) but in acid and waterlogged podzolic soils, the soil water in the mineral horizon became initially more acid and aluminium concentration rose, attributed to increased nitrification following liming and cultivation (Hornung *et al.*, 1990). This transient acidification was neutralized in later years as lime-rich water percolated down the soils (Reynolds *et al.*, 1993). Trout densities in the streams improved by liming are now reported to be similar to those of neutral streams (Donald and Gee, 1992).

Another soil liming, of riparian areas of the Cumbrian River Esk (about 10% of the catchment) at 5 t ha^{-1}, was undertaken in 1986–87 (Diamond *et al.*, 1992). Streamwater quality improved (pH and aluminium) but a similar change was also seen in the adjacent River Duddon and so was attributed to lower acid deposition (but not measured). Calcium concentrations in the treated river were initially lower post-liming, but later rose slightly; this surprising result may have arisen because of emphasis on water samples taken at high flows.

A modelling exercise based largely on these and other findings (Warfvinge and Sverdrup, 1988; Dalziel *et al.*, 1992b) suggests that the most effective lime to apply is finely particulate, that application should exceed the exchangeable acidity in the top 5 cm of soil, and that 50% of the discharge area active at high flow rates should be treated.

Of course, catchment or water-source liming may influence both terrestrial and aquatic systems, both replenishing soil base reserve and providing a buffered water quality environment, though possibly not favouring acidophilic species. However, adverse effects from terrestrial liming or from direct applications to lakes and streams seem relatively slight, and largely speculative; at Loch Fleet these seem limited to loss of one of five *Sphagnum* species directly exposed to lime in a limited wetland source area (3% of 110 ha), and on moorland to fewer female pigmy shrews, although

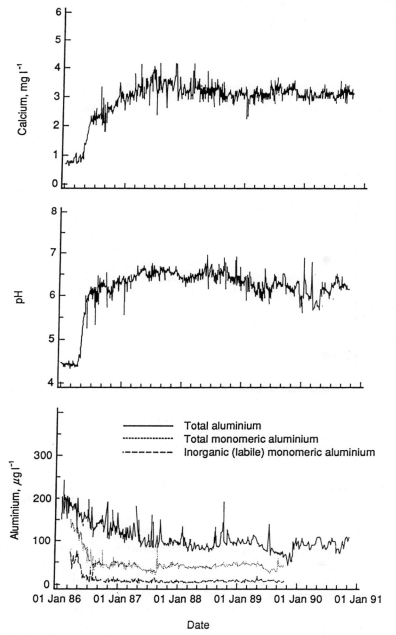

Fig. 11.11—Changes in pH, calcium and aluminium concentrations at Loch Fleet outlet following liming in 1986 (after Howells *et al.*, 1993).

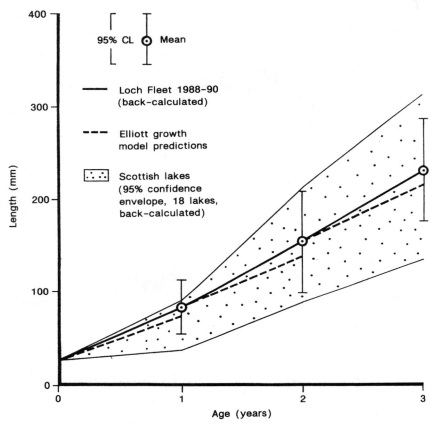

Fig. 11.12—Growth of brown trout in Loch Fleet in four years following liming, compared with data for other Scottish lakes (after Howells *et al.*, 1992).

not of males or common shrews (Shore and Mackenzie, 1993), consistent with findings over a shorter period (three years) in small experimentally limed plots in Wales.

Other methods

Other chemical manipulations used to counter acidity in lakes or rivers include the use of sodium hydroxide, sea water (Skogheim *et al.*, 1986), and nutrients (Davison, 1987, 1990) which encourage reduction of sulphate in sediments. Possible hydrological options could be diversion or dilution of acid waters by more alkaline ground waters (Cook *et al.*, 1990; Garrison *et al.*, 1991), or the longer retention of acid waters so as to optimize sediment accumulation and in-lake ANC production.

Additional land-use/management strategies have been explored to assess their potential to counter acidification of both soil and surface waters. These could include replacement of highly acidifying coniferous trees by broadleaf species, forest cutback at least in riparian zones, and ploughing regimes (Ormerod *et al.*, 1990; Welsh and

Burns, 1987) or burning of long-standing moorland vegetation (Howells and Dalziel, 1992). However, these strategies seem less effective than liming.

It is also evident that the sensitivity of species or strains provides an alternative approach; more tolerant North American brook trout introduced to Norwegian lakes have replaced lost fisheries in some. The highly sensitive rainbow trout, introduced into Europe in 1882, has become a popular stock or food fish, but has seldom bred successfully and is not suitable for acid waters, or those susceptible to short-term episodes. Introduction of more sensitive but desirable game fish to Adirondack waters in the past undoubtedly played some part in the recent loss of populations there. Similarly, fish introduced to Lake Gardsjon between 1910 and 1938 (roach, tench and carp) were unable to tolerate worsening acid conditions there (Hultberg, 1985). Restocking of Loch Fleet demonstrated the advantages of tolerant versus sensitive strains (Turnpenny et al., 1988) and some Norwegian strains of brown trout have shown a wide range of tolerance (Dalziel and Lynam, 1992; Rosseland et al., 1990). A practical fisheries management option is thus to select strains for acid-sensitive waters for such tolerance, rather than for their growth or 'fighting' characteristics.

11.8   UNRESOLVED ISSUES

Field experiments and observations are reasonably consistent with predictions of acidification models of acidification and recovery in some instances. Apparent inconsistencies may be due to different site conditions, inadequate data, or lack of understanding. The principal uncertainties are as follows.

— The contribution of sources of acid deposition, and the lack of predictable site-specific relationships, so that the outcome of emission reduction is uncertain.
— The notable heterogeneity of soils and hydrological pathways, so that 'lumped parameter' models do not predict responses to change for specific affected sites within a sensitive area.
— The response of natural soils to changes in climatic and land-use practices has seldom been explored over sufficient time.
— The size and mobility of the soil S reserve, the possible changes in N cycling, and quantification of the within-soil acid generation are not known.
— The timescale over which susceptible sites/areas will become acidified, or recover with emission control, is likely to be long-term, but modifying conditions are seldom known.
— The degree to which acid episodes are enhanced by further acid input, or are reduced by emission controls, is not known.
— The timing and sequence of biological recovery in aquatic systems is known for only a few, not necessarily characteristic, examples.

11.9   SUMMARY AND CONCLUSIONS

Acid deposition has evidently brought changes to sensitive soils, and in turn to runoff and surface waters in the northern hemisphere. It follows that some acid surface waters are inhospitable to some fish and other species.

Lower levels of sulphate deposition since 1980 (or earlier) are evident in lower sulphate concentrations in surface waters, but neither rain nor surface waters have significantly lost acidity. This may be a response to higher deposition of N species, or to other phenomena.

These observations suggest that control of acid emissions is either ineffective at specific sensitive sites or insufficient. Further, it has had little effect on fishery status or other aquatic biota.

Results of experimental exclusion of acid deposition also demonstrate that residual acidification of runoff continues for at least a decade.

Only at the massively damaged Sudbury site has emission reduction led to recovery of water quality to the extent that fish populations can be restored.

Consistent with these findings, the slow response to experimental or modelled reduction in deposition is attributable to the reserve of sulphur within soils.

As a consequence, reduction in S emissions will not improve water quality in the short term—indeed most models suggest that return to neutral pH levels will require > 80% S emission control and about a century.

During this period, N species may take over the acidification role, and will be more difficult to control.

Biological recovery in affected lakes and streams seems likely to take longer and they are unlikely to return to a 'pristine' or pre-acid state.

Reduction of acidity in rain, and possibly in soil acidity, will reduce the yield of base cations in runoff and surface waters, with continued stress for fish due to low calcium.

Liming of acid waters directly, or liming catchment or water-source areas, will improve water quality in surface waters within a much shorter timescale; liming of catchment soils will provide a higher base status and relatively stable runoff conditions for substantial periods.

Adverse effects of surface water or catchment liming appear to be limited. Beneficial effects include the restoration of fish populations and some acid-sensitive invertebrates, as well as a community close to that of circumneutral waters.

Alternative or additional strategies of some promise include chemical manipulation of water quality directly, liming to increase the base saturation of soils, hydrological management of waters, land-use changes, or fisheries management.

12

The overall view

12.1 INTRODUCTION

This text has reviewed aspects of 'acid rain' and freshwater acidification, following the path from emissions, atmospheric reactions, deposition, vegetative canopy interactions, soil responses, surface water chemistry and response of the aquatic biological communities. In addition, the potential for restoration or mitigation is discussed. Throughout almost three decades of research, findings and interpretations have often been controversial; a more objective view is now possible in the light of recent research and assessment. It is for the reader to judge, on the basis of the scientific, peer-reviewed literature, the degree and extent of acidification and the need and effectiveness of its control. Some final thoughts and a summary of major findings and future research prospects are included in this chapter.

12.2 SOME GENERAL ISSUES

The effort given by European and North American science to study of acid rain and acidification has been a remarkable phenomenon. It is appropriate to consider some conclusions that could be relevant to scientific investigation of other global-scale changes, such as global warming, loss of biodiversity, or human population problems.

Initial difficulties in acid rain research arose because of over-simplified concepts of acid rain formation and its effects—only S emissions were implicated, and a direct response was expected between rain and surface waters. This is now evidently not the case, and indeed should have been anticipated from existing scientific knowledge. The evidence of an inherent *ecological* nature of such a problem should be recognized by scientists with a diversity of disciplines; ecology is not just a biological discipline.

It is inevitable that global/national issues will involve governments, which will often promote their own concerns (seldom the advance of science). From the start, national policies for control of S emissions drove scientific activities, making it difficult for objectivity to be sustained, in some cases hindering rather than encouraging scientific advances. Scientists have a useful role in advising government objectively, but need not be surprised if their insights are not accepted. International agencies (e.g. UNEP, UNECE) have an important function in finding the level of general agreement between scientists and developing a consensus view that could guide governmental policy. However, it should not be forgotten that science does not advance by consensus, but rather by provocative hypotheses and unexpected observations that challenge established theories; politicians are likely to be impatient with such steps towards a goal of 'ultimate truth'.

Nevertheless, the acid rain initiatives of OECD, UNEP and UNECE provided a framework within which European governments could encourage their national scientific establishments to work together to a common plan. This promoted parallel investigations and healthy debate regarding interpretation of findings. Collaboration between scientists of different disciplines and nationalities has been most effective.

Nature generally changes slowly and provides many compensatory checks and balances (perhaps the 'Gaia' concept is always true) and scientific programmes to observe any change may have to match its timescale. After the first exciting flush of scientific results, most scientists are bored with finding the same, or scarcely different, results, but long-term monitoring is essential to record the decline or restoration of natural ecosystems or to confirm predictions. It is difficult to convince funding agencies of this against competition from exciting new scientific advances. The benefits of a policy for emission control (including CO_2) will not be evident in the short term—but how else will we recognize its effectiveness?

The understanding of acidification has profited greatly from large-scale field experiments, such as whole-lake or catchment studies. These can provide a realistic test of hypotheses developed from the more restricted experiments in the laboratory. If they are to be relevant to current levels of pollution, however, they may need realistic timescales of observation and funding.

Thus, the story that unfolds from views coloured by nineteenth-century conditions of gross pollution in cities and near dirty industries, through the complacency developing from overcoming that gross problem to consideration of how natural ecosystems can be protected, is one of growing scientific and technical sophistication. It seems a pattern likely to be followed by pursuit of solutions to other global problems.

12.3   MAJOR RESEARCH FINDINGS

The principal findings of acid rain research, mostly relevant to the temperate northern hemisphere, are briefly summarized below. This provides a basis for selecting some of the most important unresolved issues, and for some suggestions as to what new or additional research might be relevant in today's scientific and economic climate.

Emissions and atmosphere

Acid emissions (SO_2, NO_x, NH_x, CO_2, VOCs) are subject to oxidation in atmosphere, increasing solubility and the deposition of acidity. The deposition 'loading' is dependent on their concentrations, but also on precipitation volume (seasonally and annually).

Ozone and other photochemical oxidants in the atmosphere strongly influence the rate of oxidation. Other atmospheric materials such as dusts may be neutralizing.

S and N species are the major precursors of acid rain; their dispersion and deposition can be predicted but as yet with poor accuracy.

The relative contribution of local and more distant sources of acid gases, and that from global sources and from natural sources, remains obscure. The effect of source control at specific sensitive target sites is difficult to predict.

Dry and wet deposition are the major removal mechanisms from the atmosphere; overall they are about equal. Impaction of droplets in mist and fog, and co-deposition of NH_4^+ and SO_4^{2-}, enhance deposition in some conditions.

Vegetation effects

Acid rain at current levels in Europe and North America is unlikely to affect growth and yield of crop species, forests or natural vegetation in rural areas.

Effects on sensitive targets (mosses and lichens) at upland moor sites are attributable to gaseous pollutants, and are localized rather than a generality.

Forest decline takes a variety of forms and is not a specific disease with a single cause. Climatic changes can explain their temporal coincidence in Europe. Acid rain does enhance magnesium leaching from canopies and soils, perhaps significantly in magnesium-deficient areas, especially where forests have been intensively exploited.

Tree canopies scavenge air constituents, especially SO_4^{2-} and Cl^-, increasing their deposition in throughfall. Canopy leaching is also enhanced by acid rain, but the significance for surface waters receiving runoff from forest areas is not established, since recycling of leached ions between canopy and soils is enhanced.

Soils and catchments

Bedrock geology and soils are the prime determinants of the sensitivity of a region to acid rain. Acid soils generate acid waters, but acid deposition load and hydrology are important factors. The heterogeneity of soils on a small scale explains the differing quality of adjacent waters but is an impediment to developing a regional protective policy.

Vegetative growth, especially of trees and permanent grasslands, causes soil acidification by withdrawal of base cations into the above-ground biomass, as well as promoting acid conditions in the rhizosphere. Microbial and respiratory activity in soils also enhances soil acidity in upper horizons. At long-undisturbed sites, acidity accumulates with time.

The generation and reserve of acidity within soils greatly exceeds that deposited in rain, although increasing acidification of soils has occurred over the last 100 years.

Soil-generated acids include organic as well as inorganic acids; in some peaty or podzolized sites, the former may predominate. The degree to which soil acidity is neutralized depends on the base reserve, which in turn is dependent on the rate of weathering of geological materials. This supply of base cations reflects soil acidity, as well as that from rain, affecting run-off water quality.

Retention of deposited sulphur in soils is evident, and its potential release will delay the effect of reduced deposition. A similar accumulation of nitrogen in excess of that needed for plant growth is a matter of current concern.

Soils are the origin of aluminium, ionized in acid soil conditions and released in runoff to surface waters.

Land-use practices influence the acidification of soils and surface waters; the use of fertilizers and the intensive cultivation of conifer forests leads to enhanced acidification, improved draining limits soil neutralization of runoff, and podzolization restricts access of soil waters to lower, mineral soil horizons. In contrast, other agricultural practices, including liming and ploughing of podzols, will help to counter acidification.

Surface waters

Acid surface waters are typically oligotrophic with low ionic strength and low calcium concentrations. The volume and pattern of rainfall as well as its acidity are important factors. Interactions with vegetation and soils in the catchment are critical. In some peat and bog areas, acid production exceeds that from deposition and runoff may be more acid than rain.

Acid waters are generally defined as pH < 5.0, and alkalinity (or ANC) low. Some waters are permanently acid, but others, even in areas with low acid 'loading', are transiently acid when large acid inputs follow snowmelt or heavy rains. In maritime areas, circumneutral 'sea salt rains' can mobilize Al^{3+} and H^+ from acid soils.

Sulphate is the major anion strongly linked with acid conditions, but increasing concentrations of nitrate are observed. In maritime areas, chloride may be the major anion present, but is usually matched by sodium and so considered unimportant in the acid–base balance. Sulphate in deposition is reflected by sulphate concentrations in surface waters; a similar relationship of nitrate in deposition and in surface waters is not so clear.

Acid waters often have higher trace metal concentrations than circumneutral waters. In particular, soluble inorganic aluminium species are significant agents of toxicity, possibly also in conjunction with other trace metals. Organic acids, if present, moderate aluminium toxicity.

There is evidence, mostly from the fossil diatom record in lake sediments, that surface waters have become acidified in sensitive areas in the last 150 years. However, the rate and extent of acidification is not always explained by the level of 'acid loading' or the sensitivity of catchments.

Regions in Scandinavia, northern Europe and North America are those most affected; only for the northeast USA is there a quantitative estimate of waters acidified by acid rain. These regions are characterized by soils poor in base reserves, by high levels of acid emissions and high rainfall.

Acid waters and fisheries

Fish communities in acid waters are restricted to species tolerant of low pH and associated water quality. Salmonids and a few cyprinid species are the most sensitive.

Acid exposure of fish leads to impairment of the gill membrane function responsible for maintaining the ion balance of blood and tissues. If aluminium is present, respiratory function may also be impaired.

In northern Europe and North America, increasing acidity of surface waters is associated with the loss of some fish populations over the last 50 years; some waters of pH < 5.0 lack fish.

Other water quality characteristics, as well as acidity, are important; sufficient calcium to protect gill membrane function seems to be the most critical. In addition, the presence of toxic inorganic aluminium, and the ameliorating presence of organic acids, silicon or fluoride which complex aluminium are significant.

Other aquatic biology

The flora and fauna of acid waters is less diverse than those of circumneutral waters; some species needing a sufficient level of calcium, such as Mollusca and some Crustacea, are often absent.

In contrast, the reduced competition may make room for some opportunistic species; these may be regarded as 'nuisance' species.

In the absence of fish predation, some communities of acid waters, e.g. corixids and coleopterans, may flourish.

Basic biological processes, degradation, N-cycling, respiration and primary production, appear unaffected.

Other characteristics of upland waters, such as altitude, lake and stream size, riparian and littoral vegetation, and physical conditions such as flow and substrate, are important determinants of the aquatic biota.

Effects on riparian species of birds, amphibia and mammals are limited to a few examples.

Reversibility or mitigation

Reduced S emissions have led to some fall in surface water sulphate concentrations, but neither rain nor surface water acidity has significantly changed.

This may be explained in terms of a parallel increase in N emissions, or in the lag in reducing the soil reserve of sulphur and acidity, or the progressive decline in base reserve of soils.

Reduction of the acid input from atmosphere has lowered the supply of base cations weathered in soils. This limits the benefit to fish in oligotrophic waters, where a sufficient calcium concentration is critical.

Field experiments and modelling also suggest that recovery of surface waters to circumneutral conditions will be slow—even a complete ban on sulphur emissions will require about a century for a reversal to earlier conditions.

Alternative mitigative options include direct surface water and catchment liming, hydrological management, alternative land-uses, and fisheries management. Liming provides the quickest and most cost-effective treatment; adverse effects are few and can be limited or avoided. It can be argued that liming is a necessary adjunct to emission control, since it provides the calcium necessary for fish well-being.

12.4 UNANSWERED QUESTIONS

Although some of the unanswered questions identified in 1990 have indeed been resolved, it is inevitable that research has thrown up new questions, or has not yet been able to resolve others. Some of the most pressing issues are listed below.

- **Atmosphere**
    — The importance and quantitative significance of atmospheric oxidants;
    — the role of dusts in affecting climate and atmospheric acid generation on a long-term basis;
    — the validity and precision of dispersion models for both S and N species in relation to the scale of sensitive target sites;
    — the relative contribution of global, regional and local sources, including natural sources.

- **Deposition**
    — The significance of seasonal and short-term patterns of deposition to soil and surface water acidification;
    — the importance of enhanced deposition through mist/fog and co-deposition at sensitive sites;
    — the relative importance of sulphur and nitrogen deposition, their sources and transformations.

- **Soils and catchments**
    — The significance of soil reserves in alkalinity, acidity, base reserve, accumulated S and N;
    — short-term responses to deposition episodes, including 'sea salt' rains;
    — quantification of nitrification and denitrification processes for a variety of soil conditions.

- **Surface waters**
    — Relationship of surface water quality to changes in emissions and deposition;
    — further analysis of the dynamic equilibria of aluminium species, and complexes with other water components;
    — fishery response to acidification and mitigation on a population and community basis;
    — other aquatic biology responses and timescales and targets of recovery.

12.5   WHAT NEW RESEARCH IS NEEDED?

The focus of scientific and political interest may now be directed at other global problems, but there are still challenging scientific problems that need attention, if only because of their relevance to other issues of current concern. While an almost endless list of research possibilities often grows from new knowledge, a few selected items seem those most worth pursuing:

— nitrogen cycles, quantification of nitrification and denitrification in both terrestrial and aquatic ecosystems;
— conditions in a variety of other acid environments such as Amazon tributaries, volcanic or glacier lakes, and high-conductivity sand pit pools;
— the generation and chemical transformations of organic acids in soils;

— the further understanding of aquatic invertebrate communities; response to conditions other than acidity, although often associated;

— effects on a broad enough spatial scale to quantify the degree to which surface waters are acidified (*cf.* the NAPAP northeast USA survey);

— the timescale of recovery with alternative scenarios of S and N emission control; whether, indeed, a return to previous conditions *can* happen.

12.6   ENVOI

Much has been achieved since 1970, when acid rain was the major scientific challenge—large data sets have been accumulated, significant processes revealed and quantified, and mechanisms of toxicity identified. We have now a considerable understanding of acidification, the factors that influence its extent and degree, and its observed effects. This has led to the development of predictive models that can foresee the benefits of a variety of emission scenarios, or alternatively 'hindcast' to unreported conditions in the past. They provide a measure against which the benefits and costs of alternative mitigative options can be measured. It has become very clear, both from observations through the progressive reduction of S emissions in Europe and North America, and from the predictions of models, that reversal of the effects of acid rain cannot be achieved by emission control alone, however worthy that aim. Some replenishment of the base reserve of soils will be required in addition, and also perhaps some positive management of ecosystems to maintain the qualities considered desirable. Not all possibilities have been tested or are effective and economic in realistic conditions, and further ideas and trials are needed. As always, the task of science is never done, although politicians and governments will always say 'enough is enough' and deem it sufficient for legislation and action.

References

Abrahamsen G. (1984) Effects of acidic deposition on forest soil and vegetation. *Phil. Trans. Roy. Soc. London* **305B**: 369–382.

Abrahamsen G., K. Bjor, R. Horntvedt, B. Tveite (1976) Effects of acid precipitation on coniferous forests, pp. 36–63 in *Impact of Acidic Precipitation on Forests and Freshwater Ecosystems in Norway*, ed. F. H. Brakke, SNSF Report FR6/76.

Abrahamsen G., A. Stuanes, K. Bjor (1979) Interaction between simulated rain and barren rock. *Water Air Soil Pollut.* **11**: 191–200.

Adams W. A., A. Y. Ali, P. J. Lewis (1990) Release of cationic aluminium from acidic soils into drainage waters and relationship with land use. *J. Soil Science* **41**: 255–268.

Agren G. I. and E. Bosatta (1988) Nitrogen saturation of terrestrial ecosystems. *Environ. Pollut.* **54**: 185–197.

Ahl T. (1986) Assessment and monitoring of acidification in rivers and lakes in Sweden. UNECE Workshop, Grafenau, FRG, 28–30 April 1986.

Alabaster J. S. and R. Lloyd (1982) *Water Quality Criteria for Freshwater Fish*, Butterworths, London, 297 pp.

Almer B. and M. Hanson (1980) *Forsurningseffeker i Vastkustsjoar*. Inform. Drottningholm, 44 pp.

Almer B., W. Dickson, C. Ekstrom, E. Hornstrom, U. Miller (1974) Effects of acidification in Swedish lakes. *Ambio* **3**: 30–36.

Almer B., W. Dickson, C. Ekstrom, E. Hornstrom (1978) Sulfur pollution and the aquatic ecosystem, pp. 273–311 in *Sulfur in the Environment: Part II Ecological Impacts*, ed. J. O. Nriagu, John Wiley and Sons, New York.

Altshuller A. P. and R. A. Linthurst (1984) *The Acidic Deposition Phenomenon and its Effects*, Vol. 1, EPA-600/8-83, Washington, DC.

Andersen R., I. P. Muniz, J. Skurdal (1984) Effects of acidification on age class composition of arctic char (*Salvelinus alpinus* L.) and brown trout (*Salmo trutta* L.) in a coastal area, southwest Norway. *Freshwat. Res. Drottningholm* **61**: 5–15.

Anderson N. J., R. W. Battarbee, P. G. Appleby, A. C. Stevenson, F. Oldfield, J.

Darley, G. Glover (1986) Palaeolimnological evidence for the recent acidification of Loch Fleet, Galloway. Report 17, Palaeolimnological Res. Unit, University College London, 71 pp.

Andersson F. and B. Olsson (editors) (1985) *Lake Gardsjon: An Acid Forest Lake and its Catchment*, Swedish Res. Councils, *Ecol. Bull.* **37**: 336 pp.

Andersson G. (1985) Decomposition of alder leaves in acid lake waters. *Ecol. Bull.* **37**: 293–300.

Andersson P. and P. Nyberg (1984) Experiments with brown trout, *Salmo trutta*, in mountain streams at low pH and elevated levels of iron, manganese and aluminium. *Inst. Freshwat. Res. Drottningholm* **61**: 34–47.

Armentano T. V. and J. P. Bennett (1992) Air pollution effects on the diversity and structure of communities, pp. 159–176 in *Air Pollution Effects on Biodiversity*, ed. J. R. Barker and D. T. Tingey, Van Nostrand Reinhold, New York.

Ashmore M. R., J. N. Bell, A. J. Rutter (1988) Effects of acid rain on trees and higher plants, pp. 39–54 in *Acid Rain and Britain's Natural Ecosystems*, ed. M. R. Ashmore, N. J. Bell, C. Garretty, Imperial College Centre Environ. Technol.

AWRG (Acid Waters Review Group) (1987) Analysis of synoptic chemical data for United Kingdom surface waters using the Henriksen empirical technique. Report to UK Dept. Environ., 20 pp.

Bache B. W. (1982) The role of soil in determining water composition, pp. 434–150 in *Coal Fired Power Stations and the Aquatic Environment*, Water Chemistry Institute, Copenhagen.

Bache B. W. (1984) Soil–water interactions. *Phil. Trans. Roy. Soc. London* **305B**: 135–150.

Bache B. W. (1993) Acid rain on soils—does it matter? *Seesoil* **8**: 43–52.

Baker J. P., D. P. Bernard, S. W. Christensen, M. J. Sale (1990) Biological effects of changes in surface water acid–base chemistry. *Acid Deposition: State of Science and Technology*, NAPAP Report 13, Govt. Printing Office, Washington, DC, 381 pp.

Baker J. P. and C. L. Schofield (1982) Aluminium toxicity to fish in acidic waters. *Water Air Soil Pollut.* **18**: 289–309.

Baker L. A., P. R. Kaufmann, A. T. Herlihy, J. M. Eilers (1990) Current status of surface water acid–base chemistry. *Acid Deposition: State of Science and Technology*, NAPAP Report 9, Govt. Printing Office, Washington, DC, 367 pp.

Barrett C. F., D. H. Atkins, J. N. Cape, J. Crabtree, T. D. Davies, R. G. Derwent, B. E. A. Fisher, D. Fowler, A. S. Kallend, A. Martin, R. A. Scriven, J. G. Irwin (1987) *Acid Deposition in the United Kingdom*. Second Report of the UK Acid Rain Review Group, Dept. Environ., London, 101 pp.

Barrett C. F., D. H. F. Atkins, J. N. Cape, D. Fowler, J. G. Irwin, A. S. Kallend, A. Martin, J. I. Pitman, R. A. Scriven, A. F. Tuck (1983) *Acid Deposition in the United Kingdom*, Warren Spring Laboratory, Stevenage, Herts, 72 pp.

Battarbee R. W. (1990) The causes of lake acidification, with special reference to the role of acid deposition. *Phil. Trans. Roy. Soc. London* **327B**: 339–347.

Battarbee R. W., N. J. Anderson, P. G. Appleby, R. J. Flower, S. C. Fritz, E. Y. Haworth, S. Higgit, V. J. Jones, A. Kreiser, M. A. R. Munro, J. Natkanski, F.

Oldfield, S. T. Patrick, N. G. Richardson, B. Rippey, A. C. Stevenson (1988b) *Lake Acidification in the United Kingdom 1800–1986.* Report to Dept. Environ., University College London, 68 pp.

Battarbee R. W., R. J. Flower, A. C. Stevenson, V. J. Jones, R. Harriman, P. G. Appleby (1988a) Diatom and chemical evidence for reversibility of acidification of Scottish lochs. *Nature* **332**: 530–532.

Battarbee R. W., N. A. Logan, K. J. Murphy, P. Raven, R. J. Aston, G. N. Foster (1992) Other aquatic biology: fauna and flora, pp. 289–330 in *Restoring Acid Waters*, ed. G. Howells and T. R. K. Dalziel, Elsevier, London.

Battarbee R. W., B. J. Mason, I. Renberg, J. F. Talling (editors) (1990) *Palaeolimnology and Lake Acidification*, Royal Society, London, 445 pp.

Battarbee R. W. and I. Renberg (1990) The Surface Water Acidification Project (SWAP) palaeolimnology programme. *Phil. Trans. Roy. Soc. London* **327B**: 227–232.

Battarbee R. W., A. C. Stevenson, R. Rippey, C. Fletcher, J. Natkanski, M. Wik, R. J. Flower (1989) Causes of lake acidification in Galloway, south west Scotland: a palaeoecological evaluation of the relative roles of atmospheric contamination and catchment change for two acidified sites with non-afforested catchments. *J. Ecol.* **77**: 651–672.

Bayley S. E., R. S. Behr, C. A. Kelly (1986) Retention and release of S from a freshwater wetland. *Water Air Soil Pollut.* **31**: 101–111.

Beamish R. J. (1976) Acidification of lakes in Canada by acid precipitation and the resulting effect on fishes. *Water Air Soil Pollut.* **6**: 501–514.

Beebee T. J. C. and J. R. Griffin (1977) A preliminary investigation into the natterjack toad (*Bufo calamita*) breeding site characteristics in Britain. *J. Zool.* **181**: 341–350.

Beggs G. L. and J. M. Gunn (1986) Response of lake trout (*Salvelinus namaycush*) and brook trout (*Salvelinus fontinalis*) to surface water acidification in Ontario. *Water Air Soil Pollut.* **30**: 711–717.

Bellemakers M. J. S. and H. van Dam (1992) Improvement of breeding success of the moor frog, *Rana arvalis*, by liming of acid moorland pools and the consequences of liming for water chemistry and diatoms. *Environ. Pollut.* **78**: 165–1171.

Bernard D. P. (1990) Effects of acidification on microbial communities, pp. 66–74 in *Acidic Deposition: State of Science and Technology*, ed. J. P. Baker *et al.*, NAPAP Report SOST 13.

Birchall J. D., C. Exley, J. S. Chappell, M. J. Phillips (1989) Acute toxicity of aluminium to fish eliminated in silicon-rich acid waters. *Nature* **338**: 146–148.

Bird S. C., S. J. Brown, E. Vaughan (1990) The influence of land management on stream water chemistry, pp. 241–253 in *Acid Waters in Wales*, ed. R. W. Edwards *et al.*, Kluwer, Dordrecht, Netherlands.

Birks H. J. B., S. Juggins, J. M. Line (1990a) Lake surface-water chemistry reconstructions from palaeolimnological data. *Phil. Trans. Roy. Soc. London* **327B**: 301–313.

Birks H. J. B., J. M. Line, S. Juggins, A. C. Stevenson, C. F. F. ter Braak (1990b) Diatoms and pH reconstruction. *Phil. Trans. Roy. Soc. London* **327B**: 263–278.

Bjarnborg B. (1983) Dilution and acidification effects during the spring flood of four

Swedish mountain brooks. *Hydrobiologia* **101**: 19–26.

Blake F. (1991) A skeptical observer. *EPA Journal* **17**: 59–60.

Blank L. W. (1985) A new type of forest decline in Germany. *Nature* **314**: 311–314.

Blank L. W., H. D. Payer, T. Pfirrmann, G. Gnatz, M. Kloos, K.-H. Runkel, W. Schmolke, D. Strube (1990) Effects of ozone, acid mist and soil characteristics on clonal Norway spruce (*Picea abies* (L.) Karst.)—an introduction to the joint 14 month tree exposure experiment in closed chambers. *Environ. Pollut.* **64**: 189–208 (and following papers in this issue).

Blank L. W., T. M. Roberts, R. A. Skeffington (1988) New perspectives on forest decline. *Nature* **336**: 27–30.

Bohm-Tuchy E. (1960) Plasmalemma and the action of aluminium salts. *Protoplasma* **52**: 108.

Booth C. E., D. G. McDonald, B. P. Simons, C. M. Wood (1988) Effects of aluminium and low pH on net ion fluxes and ion balance in the brook trout (*Salvelinus fontinalis*). *Can. J. Fish. Aquat. Sci.* **45**: 1563–1574.

Boylen C. W., M. D. Schick, D. A. Roberts, R. Singer (1983) Microbiological survey of Adirondack lakes with various pH values. *Appl. Environ. Microbiol.* **45**: 1538–1544.

Bredemeier M. (1988) Forest canopy transformation of atmospheric deposition. *Water Air Soil Pollut.* **40**: 121–138.

Brett M. T. (1989) Zooplankton communities and acidification processes (a review). *Water Air Soil Pollut.* **44**: 387–414.

Brezonik P. L., L. A. Baker, J. R. Eaton, T. M. Frost, P. Garrison, T. K. Krantz, J. J. Magnuson, W. J. Rose, B. K. Shephard, W. A. Swenson, C. J. Watras, K. E. Webster (1986) Experimental acidification of Little Rock Lake, Wisconsin. *Water Air Soil Pollut.* **31**: 115–121.

Brezonik P. L., C. L. Sampson, E. P. Weir (1991) Effects of acidification on chemical composition and chemical cycles in a seepage lake: mechanistic inferences from a whole-lake experiment. Abstract pp. 251–253 Amer. Chem. Soc. Symposium Environmental Chemistry of Lakes and Reservoirs, Atlanta, GA, 201st meeting.

Brimblecombe P. (1986) *Air Composition and Chemistry*, Cambridge University Press, Cambridge, 224 pp.

Brimblecombe P. and J. T. Pitman (1980) Long-term deposit at Rothamsted, England. *Tellus* **32**: 261–267.

Brimblecombe P. and D. H. Stedman (1982) Historical evidence for a dramatic increase in the nitrate component of acid rain. *Nature* **298**: 460–462.

Brocksen R. W., M. D. Marcus, H. Olem (1992) *Practical Guide to Managing Acidic Surface Waters and Their Fisheries*, Lewis Publishers, MI, 190 pp.

Brogger W. C. (1881) Notes on a contaminated snowfall. *Naturen* **5**: 47.

Brown D J. A. and S. Lynam (1981) The effect of sodium and calcium concentrations on the hatching of eggs and the survival of yolk sac fry of brown trout, *Salmo trutta* L. at low pH. *J. Fish. Biol.* **19**: 205–211.

Brown D. J. A. and K. Sadler (1981) The chemistry and fishery status of acid lakes in Norway and their relationship to European sulphur emissions. *J. Appl. Ecol.* **18**: 433–441.

Brown K. A. (1981) Biochemical activities in peat sterilised by gamma irradiation. *Soil Biol. Biochem.* **13**: 469–474.

Brown K. A. (1982) Sulphur in the environment: a review. *Environ. Pollut.* **3**: 47–80.

Brown K. A. (1985a) Acid deposition: effects of sulphuric acid at pH 3 on chemical and biochemical properties of bracken litter. *Soil Biol. Biochem.* **17**: 38–51.

Brown K. A. (1985b) Formation of organic sulphur in anaerobic peat. CEGB Report TPRD/L/2886/N85, 13 pp.

Brydges T. G. and P. W. Summers (1989) The acidifying potential of atmospheric deposition in Canada. *Water Air Soil Pollut.* **43**: 249–263.

Bukaveckas P. A. (1992) Changes in primary productivity associated with liming and reacidification of an Adirondack lake. *Environ. Pollut.* **79**: 127–133.

Bull K. R. (1991) The critical loads/levels approach to gaseous pollutant emission control. *Environ. Pollut.* **69**: 105–123.

Caines L. A., A. M. Watt, D. E. Wells (1985) The uptake and release of some trace metals by aquatic bryophytes in acidified waters in Scotland. *Environ. Pollut.* **10**: 1–18.

Cape J. N., L. J. Sheppard, D. Fowler, A. F. Harrison, J. A. Parkinson, P. Dao, I. S. Paterson (1992) Contribution of canopy leaching to sulphate deposition in a Scots pine forest. *Environ. Pollut.* **75**: 229–236.

Carrick T. R. and D. W. Sutcliffe (1983) Concentrations of major ions in streams on catchments of the River Duddon (1971–1974) and Windermere (1975–1978), English Lake District. *Freshwat. Biol. Assocn., Occ. Publ.* **21**, 170 pp.

Carriere D., D. B. Peakall, K. L. Fischer, P. Angehra (1986) Effect of dietary aluminium sulphate on reproductive success and growth of ringed turtle doves, *Streptopelia risoria. Can. J. Zool.* **64**: 1500–1505.

Casselman J. M. and H. Harvey (1976) Selective fish mortality resulting from low winter oxygen. *Verh. Internat. Verein. Limnol.* **19**: 2418–2499.

Chamier A. C. (1987) Effect of pH on microbial degradation of leaf litter in seven streams in the English Lake District. *Oecologia* **71**: 491–500.

Chang J. S. (1990) The regional acid deposition model and engineering model, in *Acid Deposition: State of Science and Technology*, NAPAP Report 4, Washington, DC.

Chanin P. R. and D. T. Jeffries (1978) The decline of the otter, *Lutra lutra*, in Britain: an analysis of hunting records and discussion of causes. *J. Linn. Soc.* **10**: 305–328.

Charles D. F., R. W. Battarbee, I. Renberg, H. van Dam, J. P. Smol (1989) Palaeological analysis of diatoms and chrysophytes for reconstructing lake acidification trends in North America and Europe, pp. 207–276 in *Acid Precipitation, Vol. 4 Soils, aquatic processes and lake acidification*, ed. S. A. Norton *et al.*, Springer-Verlag, New York.

Charles D. F. and J. P. Smol (1991) Long term perspective on chemical changes in lakes and reservoirs: interpreting the biological record in sediments, Abstract pp. 260–263, American Chemical Society Symposium, Environmental Chemistry, Atlanta, GA (and in press).

Charlson E. J., J. E. Lovelock, M. O. Andreae, S. G. Warren (1987) Oceanic phytoplankton, atmospheric sulphur albedo and climate. *Nature* **326**: 655–661.

Chen C. W., S. A. Gherini, R. J. M. Hudson, J. D. Dean (1983) The integrated lake watershed acidification study. EPRI Report EA-3221, Vol. 1, Palo Alto, CA, 214 pp.

Chester P. F. (1984) Ecological effects of deposited sulphur and nitrogen compounds—general discussion. *Phil. Trans. Roy. Soc. London* **305B**: 564–566.

Chester P. F. (1986) Acid lakes in Scandinavia—the evolution of understanding. CEGB Report TPRD/L/PFC/010/R86, 7 pp.

Christophersen N., C. Neal, R. Vogt, J. M. Esser, S. Andersen (1990b) Aluminium mobilization in soil and stream waters at three Norwegian catchments with different acid deposition and site characteristics. *Sci. Tot. Environ.* **96**: 175–188.

Christophersen N., H. M. Seip, R. F. Wright (1982) A model for streamwater chemistry at Birkenes, Norway. *Water Resources Res.* **18**: 977–996.

Christophersen N, R. D. Vogt, R. H. Anderson, R. C. Ferrier, J. D. Miller, C. Neal, H. M. Seip (1990a) Controlling mechanisms for streamwater chemistry and soil acidification at the pristine Ingabekken site in mid-Norway; some implications for acidification models. *Water Resources Res.* **26**: 59–68.

Christophersen N. and R. F. Wright (1981) Sulfate budget and a model for sulfate concentrations in stream water at Birkenes, a small forested catchment in southernmost Norway. *Water Resources Res.* **17**: 377–389.

CLAG (1991) Acid rain—critical and target load maps for the United Kingdom. Dept. Environ., London.

CLAG (1992) Critical loads for acidity for UK freshwaters. Interim report to Dept. Environ., Critical Loads Advisory Group (in prep.).

Clark K. L. (1986) Distributions of anuran populations in central Ontario relative to habitat acidity. *Water Air Soil Pollut.* **30**: 727–734.

Clark K. L. and R. J. Hall (1985) Effects of elevated hydrogen ion and aluminium concentrations on the survival of amphibian embryos and larvae. *Can. J. Zool.* **63**: 116–123.

Clark P. A., B. E. A. Fisher, R. A. Scriven (1986) The wet deposition of sulphate and its relationship to sulphur dioxide emissions. CEGB Report TPRD/L/2955/N55.

Claussen E. (1990) Acid rain: the strategy. *EPA Journal* **17**: 21–23.

Cleveland, E. Little, S. Hamilton, D. Buckler, J. Hunn (1986) Interactive toxicity of aluminium and acidity to early life stages of brook trout. *Can. J. Fish. Aquat. Sci.* **42**: 610–620.

Clymo R. S. (1984) *Sphagnum* peat bog: a naturally acid ecosystem. *Phil. Trans. Roy. Soc. London* **305B**: 487–499.

Cogbill C. V. (1977) The effect of acid precipitation on tree growth in western North America. *Water Air Soil Pollut.* **8**: 89–93.

Cogbill C. V. and G. E. Likens (1974) Acid precipitation in the northeastern United States. *Water Resources Res.* **10**: 1133–1137.

Cohen J. B. and A. G. Rushton (1909) The nature and extent of air pollution by smoke. *Nature* **81**: 468–469.

Collier K. J. and M. J. Winterbourne (1987) Faunal and chemical dynamics of some acid and alkaline New Zealand streams. *Freshwat. Biol.* **18**: 227–240.

Conway E. J. (1942) Mean geochemical data in relation to oceanic evolution. *Proc.*

Roy. Irish Acad. **488**: 119–159.

Cook M. J., W. M. Edmunds, N. S. Robins (1990) Groundwater contribution to an acid upland lake (Loch Fleet, Scotland) and the possibilities for amelioration. *J. Hydrol.* **125**: 111–128.

Cook R. P. (1983) Effects of acidic precipitation on embryonic mortality of *Ambystoma* salamanders in the Connecticut Valley of Massachusetts. *Biol. Cons.* **27**: 77–88.

Cosby B. J., G. M. Hornberger, R. F. Wright, J. N. Galloway (1986) Modelling the effects of acid deposition: control of the long-term sulfate dynamics by soil sulfate adsorption. *Water Resources Res.* **22**: 1283–1291.

Cosby B. J., R. F. Wright, G. M. Hornberger, J. N. Galloway (1985) Modelling the effects of acid deposition: estimation of long-term water quality responses in a small forested catchment. *Water Resources Res.* **21**: 1591–1601.

Cowling E. B. (1980) An historical resumé of progress in scientific and public understanding of acid precipitation and its biological consequences. SNSF Report FR 18/80, 29 pp.

Cox R. M. (1992) Air pollution effects on plant reproductive processes and possible consequences to their population ecology, pp. 131–158 in *Air Pollution Effects on Biodiversity*, ed. J. R. Barker and D. T. Tingey, Van Nostrand Reinhold, New York.

Crane A. J. and A. T. Cocks (1987) The transport, transformation and deposition of airborne emissions from power stations. *CEGB Research* **20**: 3–15.

Cresser M. and A. C. Edwards (1987) *Acidification of Freshwaters*, Cambridge University Press, Cambridge, 136 pp.

Cribben L. D. and D. D. Scacchetti (1977) Diversity in tree species in southeastern Ohio *Betula nigra* communities. *Water Air Soil Pollut.* **8**: 47–55.

Crossley A., D. B. Wilson, R. Milne (1992) Pollution in an upland environment. *Environ. Pollut.* **75**: 81–87.

Cummins C. P. (1986) Effects of aluminium and low pH on growth and development of *Rana temporaria* tadpoles. *Oecologia* **69**: 248–252.

Dahl K. (1927) The effects of acid water on trout fry. *Salmon Trout Mag.* **46**: 35–43.

Dale J., B. Freedman, J. Kerekes (1985) Acidity and associated water chemistry of amphibian habitats in Nova Scotia. *Can. J. Zool.* **63**: 97–105.

Dalziel T. R. K., A. Dickson, P. Warfvinge, M. V. Proctor (1992b) Targets and timescales of liming treatments, pp. 365–392 in *Restoring Acid Waters*, ed. G. Howells and T. R. K. Dalziel, Elsevier, London.

Dalziel T. R. K. and S. Lynam (1991) Survival and development of four strains of Norwegian trout and one strain of Scottish trout (*Salmo trutta*) exposed to different pH levels in the absence of aluminium. PowerGen Report TR/91/23052/R, 14 pp.

Dalziel T. R. K. and S. Lynam (1992) Survival and development of four strains of Norwegian trout and one strain of Scottish trout (*Salmo salar*) exposed to different concentrations of aluminium at two pH levels. PowerGen Report PT/92/3300007/R.

Dalziel T. R. K., R. Morris, D. J. A. Brown (1986) The effects of low pH, low calcium concentrations and elevated aluminium on sodium fluxes in brown trout. *Water Air Soil Pollut.* **30**: 569–577.

Dalziel T. R. K., M. V. Proctor, K. Paterson (1992a) Water quality in surface waters before and after liming, pp. 229–258 in *Restoring Acid Waters*, ed. G. Howells and T. R. K. Dalziel, Elsevier, London.

Dannevig G. (1959) The influence of precipitation on the acidity of water courses and on fish populations. *Jaeger og Fisker* **3**: 116–117.

Darrall N. M. (1989) Effect of air pollutants on physiological processes in plants. *Plant Cell Environ.* **12**: 1–30.

Darrall N. M. and A. R. McLeod (1990) A comparison of gas exchange characteristics in *Pinus sylvestris* exposed to SO$_2$ and O$_3$ in the Liphook forest fumigation experiment. Abstract, p. 414 in *Acidic Deposition; its Nature and Impacts*, Glasgow, 1990.

Davies T. D., C. E. Pierce, H. J. Robinson, S. R. Dorling (1992) Towards an assessment of the influence of climate on wet acidic deposition in Europe. *Environ. Pollut.* **75**: 111–119.

Davis R. B. and J. P. Smol (1986) The use of sedimentary remains of siliceous algae for inferring past chemistry of lake water—problems, potential and research needs, pp. 291–300 in *Diatoms and Lake Acidity*, ed. J. P. Smol *et al.*, Junk, Dordrecht, Netherlands.

Davison W. (1987) Internal element cycles affecting the long-term alkalinity status of lakes: implications for lake restoration. *Zeit. Hydrol.* **49**: 186–201.

Davison W. (1990) Treatment of acid waters by inorganic bases, fertilizers and organic material. *Trans. Inst. Min. Metall.* **99**: A153–A157.

Degerman E. and M. Appelberg (1992) The response of stream-dwelling fish to liming. *Environ. Pollut.* **78**: 149–148.

Degerman E. and B. Sers (1992) Fish assemblages in Swedish streams. *J. Freshwat. Res.* **67**: 61–71.

Dennis R. L. (1990) Evaluation of regional acid deposition models, in *Acid Deposition, State of Science and Technology*, NAPAP Report 5, Washington, DC.

Derwent R. G., A. J. Apling, M. R. Ashmore, D. J. Ball, P. A. Clark, A. T. Cocks, R. A. Cox, D. Fowler, M. Gay, R. M. Harrison, G. J. Jenkins, P. J. A. Kay, D. P. H. Laxen, A. Martin, D. McKenna, S. A. Penkett, M. L. Williams, P. T. Woods (1987) Ozone in the United Kingdom. Dept. Environ., Harwell Laboratory, 112 pp.

Derwent R. G., O. Hov, W. A. H. Osman, J. A. van Jaarsveld, F. A. A. M. de Leeuw (1989) An intercomparison of long-term atmospheric transport models: the budgets of acidifying species for the Netherlands. *Atmos. Environ.* **23**: 1893–1909.

Diamond M., D. Hirst, L. Winder, D. H. Crawshaw, R. F. Prigg (1992) The effect of liming agricultural land on the chemistry and biology of the River Esk, northwest England. *Environ. Pollut.* **78**: 187–185.

Dickson W. (1975) Char lakes south of Dalalven. *SNV Rep.* **9**: 147 pp.

Dighton J. and R. A. Skeffington (1987) Effects of artificial acid precipitation on the mycorrhizas of Scots pine seedlings. *New Phytol.* **107**: 191–202.

Dillon P. J., R. Reid, R. Girard (1986) Changes in the chemistry of lakes following reduction of SO$_2$ emissions. *Water Air Soil Pollut.* **31**: 59–65.

Dillon P. J., R. A. Reid, E. de Grosbois (1987) The rate of acidification of aquatic systems in Ontario. *Nature* **329**: 241–243.

Dillon P. J., N. D. Yan, H. H. Harvey (1984) Acidic deposition: effects on aquatic ecosystems. *CRC Crit. Rev. Environ. Control* **13**: 167–194.

Dollard G. J., M. H. Unsworth, M. J. Harvey (1983) Pollutant transfer in upland regions by occult precipitation. *Nature* **302**: 241–243.

Donald A. P. and A. S. Gee (1992) Acid waters in upland Wales: causes, effects and remedies. *Environ. Pollut.* **78**: 141–148.

Dovland H. and A. Semb (1978) Deposition and runoff of sulphate in the Tovdal river: a study of the mass balance for September 1974–August 1976. SNSF Report IR 38/78, 22 pp.

Draaijers G. P. J., R. Van Eck, J.-E. Hallgren (1992) Relation between estimated dry deposition and throughfall in a coniferous forest exposed to controlled levels of SO_2 and NO_2. *Environ. Pollut.* **75**: 243–249.

Driscoll C. T. (1984) Procedure for the fractionation of aqueous aluminium in dilute acidic waters. *J. Environ. Anal. Chem.* **16**: 267–283.

Driscoll C. T., J. P. Baker, J. J. Bisogni, C. L. Schofield (1980) Effects of aluminium speciation on fish in dilute acidified waters. *Nature* **284**: 161–164.

Driscoll C. T., J. P. Baker, J. J. Bisogni, C. L. Schofield (1984) Aluminium speciation and equilibria in dilute acidic surface waters of the Adirondack region, in *Geologic Aspects of Acid Rain*, ed. O. P. Bricker, Proc. Amer. Chem. Soc., Ann Arbor Sciences, Ann Arbor, MI.

Driscoll C. T., R. D. Fuller, W. D. Schecher (1989a) The role of organic acids in the acidification of surface waters in the eastern U.S. *Water Air Soil Pollut.* **43**: 21–40.

Driscoll C. T., G. E. Likens, L. O. Hedin, J. S. Eaton, F. H. Bormann (1986) Changes in the chemistry of surface waters. *Environ. Sci. Technol.* **23**: 137–143.

Driscoll C. T., D. A. Schaefer, L. A. Molot, P. J. Dillon (1989b) Summary of North American data, pp. 6.1–6.45 in *The Role of Nitrogen in the Acidification of Soils and Surface Waters*, ed. J. L. Malanchuk and J. Nilsson, Nordic Council of Ministers, Miljorapport **10**.

Driscoll C. T. and W. D. Schecher (1988) Aluminium in the environment, pp. 59–120 in *Metal Ions in Biological Systems*, ed. H. Sigel and A. Sigel, Dekker, New York.

Edwards A. C., J. Creasey, U. Skiba, T. Peirson-Smith, M. S. Cresser (1985) Long term rates of acidification in UK upland acidic soils. *Soil Use Management* **1**: 61–65.

Egglishaw H., J. R. Gardiner, J. Foster (1986) Salmon catch decline and forestry in Scotland. *Scott. Geogr. Mag.* **102**: 57–61.

Egner H., G. Brodin, O. Johansson (1955) Sampling technique and chemical examination of air and precipitation. *Ann. Roy. Agric. Coll. Sweden* **22**: 369–410.

Eliassen A. (1978) The OECD study of long-range transport of air pollutants: long-range transport modelling. *Atmos. Environ.* **12**: 479–487.

Eliassen A. and J. Saltbones (1983) Modelling of long-range transport of sulphur over Europe: a two year model run and some model experiments. *Atmos. Environ.* **17**: 1457–1473.

Ellis J. C. and D. T. E. Hunt (1986) Surface water acidification: an assessment of historic water quality records. WRc Report TR 240, Marlow, Bucks, 71 pp.

EMEP (1981) Summary report of the Western Meteorological Synthesizing Centre for the first phase of EMEP, Norweg. Meteor. Inst., Blindern, Oslo.

Eriksson E. (1981) Aluminium in ground-water—the possible solution equilibria. *Nordic Hydrol.* **12**: 43–50.

Eriksson M. O. G., L. Henrikson, B.-I. Nilsson, G. Nyman, H. G. Oscarson, A. E. Stenson, K. Larsson (1980) Predator–prey relations important for the biotic changes in acidified lakes. *Ambio* **9**: 248–249.

Eriksson M. O. G., E. Hornstrom, P. Mossberg, P. Nyberg (1983) Ecological effects of lime treatment of acidified lakes and rivers. *Hydrobiologia* **101**: 145–164.

Eriksson M. O. G. and B. Tengelin (1987) Short-term effect on perch (*Perca fluviatilis*) in acidified lakes in southwest Sweden. *Hydrobiologia* **146**: 187–191.

Falkengren-Grerup U. (1986) Soil acidification and vegetation changes in deciduous forest in southern Sweden. *Oecologia* **70**: 339–347.

Falkengren-Grerup U. and H. Erikson (1990) Changes in soil, vegetation and forest yield between 1947 and 1988 in beech and oak sites in southern Sweden. *Forest Ecol. Management* (cited in Irving, 1991).

Farmer A. M. (1990) The effects of lake acidification on aquatic macrophytes—a review. *Environ. Pollut.* **65**: 219–240.

Farmer V. C. (1986) Sources and speciation of aluminium and silicon in natural waters, pp. 4–23 in *CIBA Foundation Symposium*, 121.

Fiance S. B. (1987) Effects of pH on the biology and distribution of *Ephemerella funeralis* (Ephemeroptera). *Oikos* **31**: 332–339.

Fisher B. E. A. (1983) A review of the processes and models of long-range transport of air pollutants. *Atmos. Environ.* **17**: 1865–1880.

Fjellheim A. and G. G. Raddum (1992) Recovery of acid-sensitive species of Ephemeroptera, Plecoptera and Trichoptera in River Audna after liming. *Environ. Pollut.* **78**: 173–178.

Forsberg C. and G. Morling (1987) Examples of changes in water chemistry during lake acidification and de-acidification. *Ambio* **14**: 164–166.

Forsberg C., G. Morling, R. G. Wetzel (1985) Indications of the capacity for rapid reversibility of lake acidification. *Ambio* **14**: 164–166.

Fowler D. (1992) Air pollution transport, deposition and exposure to ecosystems, pp. 31–51 in *Air Pollution Effects on Biodiversity*, ed. J. R. Barker and D. T. Tingey, Van Nostrand Reinhold, New York.

Fowler D. and P. Brimblecombe (1988) A historical perspective of acid rain in Britain, pp. 3–11 in *Acid Rain and Britain's Natural Ecosystems* ed. M. Ashmore *et al.*, Imperial College Centre Environ. Technol., London.

Fowler D., J. N. Cape, M. H. Unsworth (1989) Deposition of atmospheric pollutants on forests, *Phil. Trans. Roy. Soc. London* **324B**: 247–265.

Fowler D., J. H. Duyzer, D. D. Baldocchi (1991) Inputs of trace gases, particles and cloud droplets to terrestrial surfaces. *Proc. Roy. Soc. Edinburgh* **97B**: 35–39.

Francis A. J. (1986) The ecological effects of acid deposition. Part II. Acid rain effects on soil and aquatic microbial processes. *Experientia* **42**: 455–465.

Fraser G. A. and H. H. Harvey (1982) Elemental composition of bone from white sucker (*Catastomus commersoni*) in relation to lake acidification. *Can. J. Fish. Aquat. Sci.* **39**: 1289–1296.

Freda J. (1986) The influence of acidic pond water on amphibians: a review. *Water Air Soil Pollut.* **30**: 439–450.

Freda J. (1990) Effects of acidification on amphibians, pp. 135–151 in *Acid Deposition: State of Science and Technology* J. P. Baker *et al.* (1990) *NAPAP Report 13.*

Freda J. (1991) The effects of aluminium and other metals on amphibians. *Environ. Pollut.* **71**: 305–328.

Freda J., V. Cavdek, D. G. McDonald (1990) Role of inorganic complexation in the toxicity of aluminium to *Rana pipiens* embryos and *Bufo americanus* tadpoles. *Can. J. Fish. Aquat Sci.* **47**: 217–224.

Freda J. and W. A. Dunson (1984) Sodium balance of amphibian larvae during acute and chronic exposure to low environmental pH. *Physiol. Zool.* **57**: 435–443.

Freda J. and W. A. Dunson (1985) The influence of external cation concentration on the hatching of amphibian embryos in water of low pH. *Can. J. Zool.* **63**: 2649–2656.

Frenette J.-J., Y. Richard, G. Noreau (1986) Fish responses to acidity in Quebec lakes: a review. *Water Air Soil Pollut.* **30**: 461–475.

Fricke W. and S. Beilke (1992) Indications for changing deposition patterns in central Europe. *Environ. Pollut.* **75**: 121–127.

Fritsche U. (1992) Studies on leaching from spruce twigs and beech leaves. *Environ. Pollut.* **75**: 151–257.

Fry G. S. and A. S. Cooke (1984) Acid deposition and its implication for nature conservation in Britain. *Focus on Nature Conservation* 7, NCC, Shrewsbury, 59 pp.

Fryer G. (1980) Acidity and species diversity in freshwater Crustacea. *Freshwat. Biol.* **10**: 41–45.

Fryer G. (1985) Crustacean diversity in relation to size of water bodies: some facts and problems. *Freshwat. Biol.* **15**: 347–361.

Fryer G. and O. Forshaw (1979) The freshwater Crustacea of the island of Rhum (Inner Hebrides): a faunistic and ecological survey. *J. Linn. Soc.* **11**: 333–367.

Gahnstrom G. and S. Fleischer (1985) Microbial glucose transformation in sediment from acid lakes. *Ecol. Bull.* **37**: 287–292.

Galloway J. M., G. E. Likens, W. C. Keene, J. M. Miller (1982) The composition of precipitation in remote areas of the world. *J. Geophys. Res.* **78**: 8771–8786.

Galloway J. N., S. A. Norton, M. R. Church (1983) Freshwater acidification from atmospheric deposition of sulfuric acid: a conceptual model. *Environ. Sci. Technol.* **17**: 541A.

Galloway J. N. and H. Rodhe (1991) Regional atmospheric budgets of S and N fluxes: how well can they be quantified? *Proc. Roy. Soc. Edinburgh* **97B**: 61–80.

Gardner M. (1985) The determination of pH in poorly buffered water: recommendations of a meeting held at WRc Medmenham in April 1984. WRc Report ER 911-M, Marlow, Bucks, 44 pp.

Garrison P. J., W. Rose, C. J. Watras, J. P. Hurley (1991) Mitigation of acid rain by ground-water addition. Abstract pp. 169–170, American Chemical Society Symposium Acid Rain Mitigation, Atlanta, GA, 201st Meeting.

Gascon C. and D. Planas (1986) Spring pond water chemistry and the reproduction of the wood frog, *Rana sylvatica. Can. J. Zool.* **64**: 543–550.

Gascon C., D. Planas, G. Moreau (1987) The interaction of pH, Ca and aluminium concentrations on the survival and development of wood frog (*Rana sylvatica*) eggs and tadpoles. *Ann. Soc. Zool. Roy. Belg.* **117**: 189–199.

Gensemer R. W. (1991) The effects of pH and aluminium on the growth of the acidophilic diatom *Asterionella ralfsii* var. *americana. Limnol. Oceanogr.* **36**: 123–131.

Gherini S. A., L. Mok, R. J. M. Hudson, G. F. Davis, C. W. Chen, R. A. Goldstein (1985) The ILWAS model: formulation and application. *Water Air Soil Pollut.* **26**: 425–459.

Glover G. M. (1987) Application of the Reuss–Johnson model of soil processes to basic aluminium sulphates and other aluminium minerals. CEGB Report BAP/44/87, 11 pp.

Goldsmith P., F. B. Smith, A. Tuck (1984) Atmospheric transport and deposition. *Proc. Roy. Soc. London* **305B**: 259–279.

Goldstein R., S. A. Gherini, C. W. Chen, L. Mok, R. J. M. Hudson (1984) Integrated acidification study (ILWAS): a mechanistic ecosystem analysis. *Phil. Trans. Roy. Soc. London* **305B**: 147–178.

Gorham E. (1955) On the acidity and salinity of rain. *Geochim. Cosmochim. Acta* **7**: 231–239.

Gorham E. (1958) Atmospheric pollution by hydrochloric acid. *Quart. J. Microscop. Sci.* **84**: 274–276.

Gorham E., J. K. Underwood, F. B. Martin, J. G. Ogden (1986) Natural and anthropogenic causes of lake acidification in Nova Scotia. *Nature* **324**: 451–453.

Gosner K. L. and I. H. Black (1957) The effects of acidity on the development and hatching of New Jersey frogs. *Ecology* **38**: 256–262.

Goulding K. W. T., A. E. Johnston, P. R. Poulton (1987) The effect of atmospheric deposition, especially of nitrogen, on grassland and woodland at Rothamsted Experimental Station, England, measured over more than 100 years, in *Effects of Air Pollution on Terrestrial and Aquatic Ecosystems*, EEC Symposium, May 1987, Grenoble, France.

Grahn O., H. Hultberg, L. Landner (1974) Oligotrophication—a self-accelerating process in lakes subject to excessive supply of acid substances. *Ambio* **3**: 93–94.

Granat L. (1977) The IMI network in Sweden—present equipment, methods and plans for improvement. IMI Report AC-40.

Granat L. (1978) Sulphate in precipitation as observed by the European Atmospheric Chemistry Network. *Atmos. Environ.* **12**: 413–424.

Grennfelt P. and H. Hultberg (1986) Effects of nitrogen deposition on the acidification of terrestrial and aquatic ecosystems. *Water Air Soil Pollut.* **30**: 945–963.

Grennfelt P., S. Larsson, P. Leyton, B. Olsson (1985) Atmospheric deposition in the Lake Gardsjon area. *Ecol. Bull.* **37**: 101–108.

Grimvall A., C. A. Cole, B. Allard, P. Sanden (1986) *Water Qual. Bull.* **11**: 6.

Grip H. and K. H. Bishop (1990) Chemical dynamics in an acid stream rich in dissolved organics, pp. 75–84 in *The Surface Waters Acidification Programme*, ed. B. J. Mason, Cambridge University Press, Cambridge.

Gunn J. M. and W. Keller (1990) Biological recovery of an acid lake after reductions in industrial emissions of sulphur. *Nature* **345**: 431–433.

Haines T. A. (1987) Atlantic salmon resources in the northeastern United States and the potential effects of acidification from atmospheric deposition. *Water Air Soil Pollut.* **35**: 37–48.

Haines T. A. and J. P. Baker (1986) Evidence of fish population responses to acidification in the eastern United States. *Water Air Soil Pollut.* **31**: 605–629.

Haines T. A., S. J. Pauwels, C. H. Jagoe (1986) Predicting and evaluating the effects of acidic precipitation on water chemistry and endemic fish populations in northeastern United States. US Fish and Wildlife Biol. Report **80**, 139 pp.

Hall R. J., C. T. Driscoll, G. E. Likens (1985) Physical, chemical and biological consequences of episodic aluminium additions to a stream. *Limnol. Oceanogr.* **30**: 212–220.

Hall R. J., G. E. Likens, S. B. Fiance, G. R. Hendrey (1980) Experimental acidification of a stream in Hubbard Brook Experimental Forest, New Hampshire. *Ecology* **61**: 976–989.

Hall R. T. (1987) Processes of evaporation from vegetation of the uplands of Scotland. *Trans. Roy. Soc. Edinburgh* **78B**: 327–334.

Hallbacken L. and C. O. Tamm (1986) Changes in soil acidity from 1927 to 1982 in a forest area of southwest Sweden. *Scand. J. For. Res.* **1**: 219–232.

Hansen D. A. and G. M. Hidy (1981) Review of questions regarding rain acidity data. Environmental Research and Technology Report 81-RD-100, 68 pp.

HAPRO (1985) Finnish Research Project on Acidification. Min. Environ. and Min. Agric. and Forestry Report, 8 pp.

Haramis G. M. and D. S. Chu (1987) Acid rain effects on waterfowl: use of black duck brood to assess food resources of experimentally acidified wetlands. *ICBP Tech. Publ.* **6**: 173–181.

Harriman R. and B. R. S. Morrison (1982) Ecology of streams draining forested and non-forested catchments in an area of central Scotland and subject to acid precipitation. *Hydrobiologia* **88**: 251–263.

Harriman R., B. R. S. Morrison, L. A. Caines, P. Collen, A. W. Watt (1987) Long term changes in fish populations of acid streams and lochs in Galloway, south west Scotland. *Water Air Soil Pollut.* **32**: 89–112.

Harriman R. and D. E. Wells (1987) Causes and effects of surface water acidification in Scotland. *Water Pollut. Control Fed.* **84**: 215–244.

Harter P. (1988) Acidic deposition—ecological effects. IEA Report, 196 pp.

Harvey H. H. (1975) Fish populations in a large group of acid stressed lakes. *Verh. Internat. Verein. Limnol.* **19**: 2406–2417.

Harvey H. H. (1978) Fish communities of the Manitoulin Island Lakes. *Verh. Internat. Verein. Limnol.* **20**: 2031–2038.

Harvey H. H. (1981a) Fish communities of the lakes of Bruce Peninsula. *Verh. Internat. Verein. Limnol.* **21**: 1222–1230.

Harvey H. H. (1981b) Population responses of fishes in acidified waters, pp. 227–242 in Proc. Int. Symp. *Acidic Precipitation and Fishery Impacts.*

Harvey H. H. and C. Lee (1982) Historical fisheries changes related to surface water pH changes in Canada, pp. 45–55 in *Acid Rain/Fisheries*, ed. T. A. Haines and R. E. Johnson, Amer. Fish. Soc., Bethesda, MD.

Hauhs M. and R. F. Wright (1988) Acid deposition: reversibility of soil and watershed acidification—a review. *Air Pollut. Res. Rept. (EC)* **11**: 1–42.

Havens K. E. (1990) Aluminium binding to ion exchange sites in acid-sensitive versus acid-tolerant cladocerans. *Environ. Pollut.* **64**: 133–141.

Havens K. E. (1991) Littoral zooplankton responses to acid and aluminium stress during short-term laboratory bioassays. *Environ. Pollut.* **73**: 71–84.

Havens K. E. and R. T. Heath (1989) Acid and aluminium effects on freshwater zooplankton: an *in situ* mesocosm study. *Environ. Pollut.* **62**: 195–211.

Hawkesworth D. L. and F. Rose (1970) Qualitative scale for estimating sulphur dioxide air pollution in England and Wales using epiphytic lichens. *Nature* **227**: 145–148.

Heitto J. (1990) A macrophyte survey in Finnish lakes sensitive to acidification, SIL Congress, Munich 1989 (cited in J. P. Baker *et al.*, 1990).

Helme N. and C. Neme (1990) Acid rain: the problem. *EPA Journal* **17**: 18–20.

Heltcher K. and D. Bernard (1990) Effects of acidification on macrophytes, pp. 102–113 in *Acid Deposition: State of Science and Technology* J. P. Baker *et al.* (1990) NAPAP Report SOST 13.

Heltcher K., D. Marmorek, D. Bernard (1990) Effects of acidification on zooplankton, pp. 83–91 in *Acid Deposition: State of Science and Technology* J. P. Baker *et al.* (1990) NAPAP Report SOST 13.

Henderson P. A. (1985) An approach to the prediction of temperate freshwater fish communities. *J. Fish. Biol.* **27A**: 279–291.

Henderson P. A. and I. Walker (1986) The leaf litter community of the Amazonian blackwater stream Tarumazinho. *J. Trop. Ecol.* **2**: 1–17.

Hendrey G. R., K. Baalsrud, T. S. Traaen, M. Laake, G. Raddum (1976) Acid precipitation: some hydrobiological changes. *Ambio* **5**: 224–227.

Hendrey G. R. and F. A. Vertucci (1980) Benthic plant communities in acidic Lake Golden, New York, pp. 314–315 in *Ecological Impact of Acid Precipitation*, ed. D. Drablos and A. Tollan, SNSF Report, Oslo, Norway.

Henriksen A. (1979) A simple approach for identifying and measuring acidification in fresh water. *Nature* **278**: 542–545.

Henriksen A. (1980) Acidification in freshwaters—a large scale tritration, pp. 68–74 in *Ecological Impact of Acid Precipitation*, ed. D. Drablos and A. Tollan, SNSF Report, Sandefjord, Norway.

Henriksen A. (1982a) Susceptibility of surface waters to acidification, pp. 103–121 in *Acid Rain/Fisheries*, ed. T. A. Haines and R. E. Johnson, Amer. Fish. Soc., Bethesda, MD.

Henriksen A. (1982b) Changes in base cation concentrations due to freshwater acidification. NIVA Report OF-81623, Oslo, Norway, 50 pp.

Henriksen A. (1982c) Alkalinity and acid precipitation research, *Vatten* **1**: 83–85.

Henriksen A. and D. Brakke (1988) Increasing contributions of nitrogen to the acidity of surface waters in Norway. *Water Air Soil Pollut.* **42**: 183–201.

Henriksen A., E. Joranger, B. Jonsson, B. Kvaeven (1988c) Norwegian monitoring programme for long-range transported pollutants: annual report on observations in 1987. Statens Forurensingstilsyn Report 333/88, 242 pp.

Henriksen A., J. Kamari, M. Posch, A. Wilander (1992) Critical loads of acidity: Nordic surface waters. *Ambio* **21**: 356–363.

Henriksen A., L. Lien, B. O. Rosseland, T. S. Traaen, I. V. Sevaldrud, G. Raddum, A. Fjellheim (1988b) 1000 lake fish status, Norway. Norwegian State Pollution Control Authority Report 314/88, Oslo, Norway, 36 pp.

Henriksen A., L. Lien, T. S. Traaen, I. V. Sevaldrud (1987) The 1000 lake survey, 1986, Norway. Norwegian State Pollution Control Authority Report 283/87, Oslo, Norway, 33 pp.

Henriksen A., L. Lien, T. S. Traaen, I. V. Sevaldrud, D. F. Brakke (1988a) Lake acidification in Norway—present and predicted chemical status. *Ambio* **17**: 259–266.

Henriksen A., O. K. Skogheim, B. O. Rosseland (1984) Episodic changes in pH and aluminium speciation kill fish in a Norwegian salmon river. *Vatten* **40**: 255–260.

Henriksen A., B. M. Wathne, E. J. S. Rogeberg, S. A. Norton, D. F. Brakke (1988d) The role of stream substrates in aluminium mobility and acid neutralization. *Water Res.* **22**: 1069–1073.

Hesthagen T., H. M. Berger, B. M. Larsen, T. Nost, I. V. Sevaldrud (1992) Abundance and population structure of perch (*Perca fluviatilis* L.) in some acidic Norwegian lakes. *Environ. Pollut.* **78**: 97–101.

Hicks B. B. and R. S. Artz (1992) Estimating background precipitation quality from network data. *Environ. Pollut.* **75**: 137–143.

Highton N. H. and M. J. Chadwick (1982) The effects of changing patterns of energy use on sulfur emissions and deposition in Europe. *Ambio* **11**: 324–329.

Hoeniger J. F. M. (1985) Microbial decomposition of cellulose on acidifying lakes of south-central Ontario. *Appl. Environ. Microbiol.* **50**: 315–322.

Hooper R. P. and C. A. Shoemaker (1985) Aluminium mobilization in an acidic headwater stream: temporal variation and mineral dissolution disequilibria. *Science* **229**: 463–465.

Hornung M. (1993) Critical load concepts, pp. 78–94 in *The Chemistry and Deposition of Nitrogen Species in the Troposphere*, ed. A. T. Cocks, Roy. Soc. Chem., UK.

Hornung M., S. J. Brown, A. Ranson (1990a) Amelioration of surface water acidity by catchment management, pp. 311–328 in *Acid Waters in Wales*, ed. R. W. Edwards *et al.*, Kluwer, Dordrecht, Netherlands.

Hornung M., S. Le-Grice, N. Brown, D. Norris (1990b) The role of geology and soils in controlling surface water acidity in Wales, pp. 55–66 in *Acid Waters in Wales*, ed. R. W. Edwards *et al.*, Kluwer, Dordrecht, Netherlands.

Hornung M., B. Reynolds, P. A. Stevens, S. Hughes (1990c) Water quality changes from input to stream, pp. 223–240 in *Acid Waters in Wales*, ed. R. W. Edwards *et al.*, Kluwer, Dordrecht, Netherlands.

Hornung M., F. Roda, S. J. Langan (1990d) A review of small catchment studies in western Europe producing hydrochemical budgets, Air Pollut. Res. (EC) Report **28**, 186 pp.

Howells G. (1983) Acid waters—the effect of low pH and acid associated factors on fishes. *Appl. Biol.* **9**: 143–255.

Howells G. (editor) (1994) *Water Quality for Freshwater Fish*, Gordon and Breach, Reading, 222 pp.

Howells G. and T. R. K. Dalziel (editors) (1992) *Restoring Acid Waters*, Elsevier, London, 421 pp.

Howells G., T. R. K. Dalziel, M. Proctor (1991) Nitrate and sulphate budgets at Loch Fleet, SW Scotland. Abstract, pp. 144–147, American Chemical Society, Env. Chem. Symposium *Acid Rain Mitigation*, Atlanta, GA.

Howells G. D., T. R. K. Dalziel, J. P. Reader, J. F. Solbé (1990) EIFAC water quality criteria for European freshwater fish: report on aluminium. *Chemistry and Ecology* **4**: 117–173.

Howells G., T. R. K. Dalziel, A. W. H. Turnpenny (1992) Loch Fleet: liming to restore a brown trout fishery. *Environ. Pollut.* **78**: 131–139.

Howells G., T. R. K. Dalziel, A. W. H. Turnpenny (1993) Aluminium toxicity and restoration of a brown trout fishery in a limed upland lake (Loch Fleet, Galloway, Scotland). EIFAC Symp. *Sublethal and Chronic Toxic Effects of Pollutants on Freshwater Fish*, Lugano, Switzerland, May 1992 (in press).

Huettl R. F. (1989) 'New types' of forest damages in central Europe, pp. 22–74 in *Air Pollution's Toll on Forests and Crops* ed. J. J. MacKenzie and M. T. Al-Ashry, Vail-Ballou Press, New York.

Hultberg H. (1985) Budgets of base cations, chloride, nitrogen and sulphur in the acid Lake Gardsjon catchment. *Ecol. Bull.* **37**: 133–157.

Hultberg H. and Andersson I. A. (1982) Liming of acidified lakes: induced long-term changes. *Water Air Soil Pollut.* **18**: 311–331.

Hultberg H. and O. Grahn (1980) Effects of acid precipitation on macrophytes in oligotrophic lakes. *J. Great Lakes Res.* **2**: 208–217.

Hultberg H. and P. Grennfelt (1992) Sulphur and seasalt deposition as reflected by throughfall and runoff chemistry in forested catchments. *Environ. Pollut.* **75**: 215–222.

Hultberg H. and U. Nystrom (1988) The role of hydrology in treatment duration and reacidification in the limed Lake Gardsjon, pp. 95–135 in *Liming of Lake Gardsjon*, ed. W. Dickson, SVN, Solna, Sweden.

Hunter M. L., J. J. Jones, K. W. Gibbs, J. R. Morling (1986) Duckling responses to lake acidification: do black ducks and fish compete? *Oikos* **47**: 26–32.

Hutchinson T. S. and M. Havas (1986) Recovery of some previously acidified lakes near Coniston, Canada, following reduction in atmospheric sulphur and metal emissions. *Water Air Soil Pollut.* **28**: 319–333.

Iivonen P., S. Piepponen, M. Verta (1992) Factors affecting trace metal bioaccumulation in Finnish headwater lakes. *Environ. Pollut.* **78**: 79–85.

Innes J. L. and R. C. Boswell (1987) Forest health surveys 1987: Results. UK Forestry Commission Bull. **74**, 23 pp.

Innes J. L. and R. C. Boswell (1988) Forest health surveys: Analysis and interpretation, UK Forestry Commission Bull. **88**, 52 pp.

Irving P. (editor) (1991) *Acidic Deposition: State of Science and Technology*, Summary Report of the NAPAP, Washington, DC, 265 pp.

Irwin J. G., J. N. Cape, P. A. Clark, T. D. Davies, R. G. Derwent, B. E. A. Fisher, D. Fowler, A. S. Kallend, J. W. S. Longhurst, A. Martin, F. B. Smith, D. A. Warrilow (1990) *Acid Deposition in the United Kingdom, 1986–1988*, Dept. of Environ., London, 124 pp.

Iversen T., J. Saltbones, H. Sandnes, A. Eliassen, O. Hov (1989) Airborne transboundary transport of sulphur and nitrogen over Europe. EMEP Centre, W2/89, Norweg. Meteor. Inst., Blindern, Norway.

Jacks G. (1990) Mineral weathering studies in Scandinavia, pp. 215–222 in *The Surface Waters Acidification Programme*, ed. B. J. Mason, Cambridge University Press, Cambridge.

Jacks G. and G. Knutsson (1982) Sensitivity to groundwater acidification in different parts of Sweden. KHM Tech. Report **49**, 110 pp.

Jackson S. T. and D. F. Charles (1988) Aquatic macrophytes in Adirondack (New York) lakes: patterns of species composition in relation to environment. *Can. J. Bot.* **66**: 1449–1460.

Jacobson J. S. (1984) Effects of acidic aerosol, fog, mist and rain on crops and trees. *Phil. Trans. Roy. Soc. London* **305B**: 327–333.

Jagoe C. H., T. A. Haines, F. W. Kircheis (1987) Abnormal gill development in Atlantic salmon (*Salmo salar*) fry exposed to aluminium at low pH. *Ann. Soc. Roy. Zool. Belg.* **117**: 375–386.

Jenkins D. and W. M. Shearer (1986) *The Status of Atlantic Salmon in Scotland*, NERC, Inst. Terr. Ecol. No. 15, 127 pp.

Jenkinson D. S. (1970) The accumulation of organic matter in soil left undisturbed. *Rothamsted Rep.* 1970 (2): 113–137.

Johannes A. H., E. R. Altwicker, N. L. Clesceri (1984) Atmospheric inputs to the ILWAS lake watersheds, pp. 2.1–2.17 in *The Integrated Lake–Watershed Acidification Study* **4**: EPRI Report EA-3221.

Johannessen M. and A. Henriksen (1987) Chemistry of snowmelt water. Changes in concentrations during melting. *Water Res.* **14**: 615–619.

Johannessen M., A. Skartveit, R. F. Wright (1980) Streamwater chemistry before, during and after snowmelt, pp. 224–225, in *Ecological Effects of Acid Precipitation*, ed. D. Drablos and A. Tollan, SNSF Report, Sandefjord, Norway.

Johnson A. H. (1979) Acidification of headwater streams in New Jersey. *J. Environ. Qual.* **8**: 383–386.

Johnson D. W. (1987) A discussion on changes in soil acidity due to natural processes and acid deposition, pp. 333–346 in *Effects of Acidic Deposition on Forests, Wetlands and Agricultural Ecosystems*, ed. T. C. Huchinson and K. Meem, Springer-Verlag, New York and Toronto.

Johnson D. W. and D. W. Cole (1980) Anion mobility in soils; relevance to nutrient transport from forest ecosystems. *Environ. Internat.* **3**: 79–90.

Johnson D. W., M. S. Cresser, A. I. Nilsson, J. Turner, B. Ulrich, D. Brinkley, D. W. Cole (1991) Soil changes in forest ecosystems: evidence for and probable causes. *Proc. Roy. Soc. Edinburgh* **97B**: 81–116.

Johnson D. W., G. S. Henderson, D. E. Todd (1981) Evidence of modern accumulation of adsorbed sulfate in an east Tennessee forested ultisol. *Soil Science* **132**: 449–461.

Johnson D. W., D. D. Richter, G. M. Lovett, S. E. Lindburg (1985) The effects of atmospheric deposition on potassium, calcium, and magnesium cycling in two deciduous forests. *Can. J. For. Res.* **15**: 773–782.

Johnson D. W., J. Turner, J. M. Kelly (1982) Effects of acid rain on forest nutrients. *Water Resources Res.* **18**: 449–461.

Johnston A. E., K. W. Goulding, P. R. Poulton (1986) Soil acidification during more than 100 years under permanent grassland and woodland at Rothamsted. *Soil Use Management* **2**: 3–10.

Jonsson B. (1977) Soil acidification by atmospheric pollution and forest growth. *Water Air Soil Pollut.* **7**: 497–501.

Jonsson B. and R. Sundberg (1972) Has acidification by atmospheric pollution caused a growth reduction in Swedish forests? Research Notes 20, Swedish Roy. Coll. Forestry.

Junge C. E. and R. T. Werby (1958) The concentration of chloride, sodium, potassium, calcium and sulfate in rain water over the United States. *Meteorology* **15**: 417–425.

Junk W. J. (1983) Ecology of swamps in the middle Amazon, Chapter 9 in: *Ecosystems of the World 4B (Mires, Swamp, Bog, Fen and Moor)*, Elsevier, New York.

Kallend A. S., A. R. M. Marsh, J. H. Pickles, M. V. Proctor (1983) Acidity of rain in Europe. *Atmos. Environ.* **17**: 127–137.

Kamari J., M. Forsius, P. Kortelainen, J. Mannio, M. Verta (1991) Finnish lake survey: present status of acidification. *Ambio* **20**: 23–27.

Keller W., J. M. Gunn, N. D. Yan (1992) Evidence of biological recovery in acid-stressed lakes near Sudbury, Canada. *Environ. Pollut.* **78**: 79–85.

Keller W., J. R. Pitblado, N. I. Conroy (1986) Water quality improvements in the Sudbury, Ontario, Canada area related to reduced sulphur emissions. *Water Air Soil Pollut.* **31**: 765–774.

Kelly C. A., J. W. M. Rudd, D. W. Schindler (1990) Lake acidification by nitric acid: future considerations. *Water Air Soil Pollut.* **50**: 49–61.

Kelly M. (1988) *Mining and the Freshwater Environment*, Elsevier, London, 231 pp.

Kenk G. and H. Fischer (1988) Evidence from nitrogen fertilisation in the forests of Germany. *Environ. Pollut.* **54**: 199–218.

Kentammies K. K. (1991) Liming of acid lakes in Finland—a national review, Abstract, pp. 173–174 American Chemical Society Symposium *Acid Rain Mitigation*, Atlanta, GA.

Kerekes J., G. Howell, T. Pollock (1984) Problems associated with sulphate determination in colored humic waters in Kejimkujik National Park (Canada). *Verh. Internat. Verein. Limnol.* **22**: 1811–1817.

Kleissen F. M., H. S. Wheater, M. B. Beck, R. Harriman (1990) Conservative mixing of water sources: analysis of the behaviour of the Allt a'Mharcaidh catchment. *J. Hydrol.* **116**: 365–374.

Kortelainen P. and J. Mannio (1988) Natural and anthropogenic acidity sources for Finnish Lakes. *Water Air Soil Pollut.* **42**: 341–352.

Krahl-Urban B., H. E. Papke, K. Peters, C. Schimansky (1988) *Forest Decline*, Julich Nucl. Res. Center and US EPA, KFA Julich GmbH, 137 pp.

Kramer J., C. S. Cronan, J. V. DePinto, J. F. Hemond, E. M. Pardue, S. Visser (1989) Organic acids and acidification in surface waters. Report to Acid Dep. Comm., Utility Air Regulatory Group.

Kramer J. and A. Tessier (1982) Acidification of aquatic systems; a critique of chemical approaches. *Environ. Sci. Technol.* **16**: 606A–615A.

Kretser W., J. Gallagher, J. Nicolette (1989) Adirondack Lakes Study 1984–1987: an evaluation of fish communities and water chemistry. Adirondack Lakes Survey Corporation, Ray Brook, NY, 345 pp.

Krug E. C. (1988) Acidification of Norwegian lakes. *Nature* **334**: 571.

Krug E. C. (1989) Assessment of the theory and hypothesis of acidification of watersheds. US Dept. Energy, Contract Report 457, April 1989.

Krug E. C. and C. R. Frink (1983) Acid rain on acid soil: a new perspective. *Science* **221**: 520–525.

Kullberg A. (1992) Benthic macroinvertebrate community structure in 20 streams of varying pH and humic content. *Environ. Pollut.* **78**: 103–106.

Kullberg A. and R. C. Petersen (1987) Dissolved organic carbon, seston and macroinvertebrate drift in an acidified and a limed humic stream. *Freshwat. Biol.* **17**: 553–564.

Laake M. (1976) Effect of low pH on production, breakdown, and nutrient cycling in the littoral zone. Results of field experiments in Tovdal, 1974–75. SNSF Report IR/29/76, 75 pp.

Laaksonen R. and V. Malin (1984) Changes in water quality in Finnish lakes 1965–1982. *Water Res. Inst.* **57**: 52–58.

Lacroix G. L. (1992) Mitigation of low stream pH and its effects on salmonids. *Environ. Pollut.* **78**: 157–164.

Lacroix G. L. and K. T. Kan (1986) Speciation of aluminium in acidic rivers of Nova Scotia supporting Atlantic salmon: a methodological examination. *Can. Tech. Rept. Fish. Aqat. Sci.* No. 1501, 12 pp.

Lacroix G. L. and D. R. Townsend (1987) Responses of juvenile Atlantic salmon (*Salmo salar*) to episodic increases in acidity in Nova Scotia rivers. *Can. J. Fish. Aquat. Sci.* **44**: 1475–1484.

Langan S. (1987) Episodic acidification of streams at Loch Dee, SW Scotland. *Trans. Roy. Soc. Edinburgh* **78B**: 393–397.

Langan S. (1989) Seasalt induced streamwater acidification. *Hydrol. Processes* **3**: 25–41.

Langan S. J. and R. Harriman (1993) Critical loads as a tool for acidification management, pp. 137–145 in *Acidification, Forestry and Fisheries Management in Upland Galloway*, Proc. Loch Dee Symposium, 8–9 December 1992, FWR Report FR/SC0003.

Langdon R. W. (1985) Fisheries status in relation to acidity in selected Vermont streams. Report to NCSU Acid Deposition Program, State of Vermont. Agency of Environ. Conserv., Dept. Water Resources and Environ. Engin., Montpelier, VT.

Lappalainen A., M. Rask, P. J. Vuorinen (1988) Acidification affects perch, *Perca fluviatilus*, populations in small lakes of southern Finland. *Biol. Fish.* **21**: 231–239.

Lee J. A., M. C. Press, C. Studholme, S. J. Woodin (1988) Effects of acid deposition on wetlands, pp. 27–38 in *Acid Rain and Britain's Natural Ecosystems*, ed. M. R. Ashmore, N. J. Bell, C. Garretty, Imperial College Centre Environ. Technol.

Lees F. M. and D. Farley (1993) Characteristics and trends in rainfall and surface water chemistry at Loch Dee from 1980 to 1990, pp. 23–37 in *Acidification, Forestry and Fisheries Management in Upland Galloway*, Proc. Loch Dee Symposium, 8–9 December 1992, FWR Report FR/SC003.

Lees F. M., D. J. Tervet, J. C. Burns (1989) A study of catchment acidification. Interim Report 1980 to 1986, SDD ARD Report Series.

Leino R. L., P. Wilkinson, J. G. Anderson (1987) Histopathological changes in the gills of pearl dace, *Semotilus margarita*, and fathead minnow, *Pimephales promelas*, from experimentally acidified Canadian lakes. *Can. J. Fish. Aquat. Sci.* **44**: 126–134.

Leuven R. S. E. W., C. den Hartog, M. M. C. Christiaans, W. H. C. Heijligers (1986) Effects of water acidification on the distribution pattern and reproductive success of amphibians. *Experientia* **42**: 495–503.

Leuven R. S. E. W., S. E. Wendelar Bonga, F. G. F. Oyen, W. Hagemeijer (1987) Effects of acid stress on the distribution and reproductive success of freshwater fish in Dutch soft waters. *Ann. Soc. Roy. Zool. Belg.* **117**: 231–242.

Lewis W. M. and M. C. Grant (1981) Effect of the May–June Mount St. Helens eruptions on the precipitation chemistry, central Colorado. *Atmos. Environ.* **15**: 1539–1542.

Li Y.-H. (1990) Seasalt and pollution inputs over the continental United States. *Water Air Soil Pollut.* **64**: 561–573.

Liimatainen V. A., E. J. Snucins, J. M. Gunn (1987) Observations of lake trout (*Salvelinus namaycush*) spawning behavior in low pH lakes near Sudbury, Ontario. Ontario Fisheries Acidification Report Series No. 87-10, Ont. Min. Nat. Resources.

Likens G. E., R. F. Wright, J. N. Galloway, N. J. Butler (1979) Acid rain. *Scientific American* **241**: 43–51.

Lines R. (1979) Airborne pollutant damage to vegetation—observed damage, pp. 234–241 in *Sulphur Emissions and the Environment*, Society of Chemical Industry, London.

Linzon S. N. and P. J. Temple (1980) Soil resampling and pH measurement after an 18 year period in Ontario, pp. 176–177 in *Ecological Impact of Acid Precipitation*, ed. D. Drablos and A. Tollan, SNSF Report, Oslo.

Looney J. H. H. and P. W. James (1988) Effects on lichens, pp. 13–26 in *Acid Rain and Britain's Natural Ecosystems*, ed. M. R. Ashmore, N. J. Bell, C. Garretty, Imperial College Centre Environ. Technol.

Lydersen E. (1990) The solubility and hydrolysis of aqueous aluminium hydroxides in dilute freshwaters at different temperatures. *Nordic Hydrol.* **21**: 195–204.

Lydersen E., B. Salbu, A. B. S. Poleo, I. P. Muniz (1990) The influences of temperature on aqueous aluminium chemistry. *Water Air Soil Pollut.* **51**: 203–215.

Mackereth F. J. H., J. Heron, J. F. Talling (1989) *Water Analysis*, Freshwater Biol. Assoc., Publ. No. 36, 120 pp.

McColl R. H. S. (1975) Chemical and biological conditions in lakes of the volcanic plateau, pp. 123–139, Chapter 8, in V. H. Jolley and J. M. A. Brown, Auckland University Press/Oxford University Press, New Zealand.

McDonald D. G., J. P. Reader, T. R. K. Dalziel (1989) The combined effects of pH and trace metals on fish ionoregulation, pp. 221–242 in *Acid Toxicity and Aquatic Animals*, ed. R. Morris *et al.*, Cambridge University Press, Cambridge.

McFadden J. T. (1977) The argument supporting the reality of compensation in fish populations and a plea to let them exercise it, pp. 153–183 in *Assessing the Effects of Power Plant Induced Mortality on Fish Populations*, ed. W. van Winkle, Pergamon, New York.

McWilliams P. G. and W. T. W. Potts (1978) The effects of pH and calcium concentrations on gill potentials in the brown trout, *Salmo trutta*, from an acid river. *J. Comp. Physiol.* **126**: 277–286.

Magnuson J. J., J. P. Baker, E. J. Rahel (1984) A critical assessment of effects of acidification on fisheries. *Phil. Trans. Roy. Soc. London* **305B**: 501–516.

Maitland P. S. (1987) Fish introductions and translocations—their impact in the British Isles, pp. 57–65 in *Angling and Wildlife in Fresh Waters*, ed. P. S. Maitland and A. K. Turner, NERC, Inst. Terr. Ecol. No. 19, 84 pp.

Maitland P. S. and R. N. Campbell (1992) *Freshwater Fishes of the British Isles*, Harper Collins, London, 368 pp.

Maitland P. S., A. A. Lyle, R. N. B. Campbell (1987) *Acidification and fish in Scottish lochs.* Inst. Terr. Ecol. Report, Grange-over-Sands, UK, 71 pp.

Malley D. F. and P. S. S. Chang (1985) Effects of aluminium and acid on calcium uptake by the crayfish, *Orconectes virilis*. *Arch. Contamin. Toxicol.* **14**: 739–747.

Mason B. J. (1990) *The Surface Water Acidification Programme*, Cambridge University Press, Cambridge, 522 pp.

Mason C. F. and S. M. MacDonald (1987) Acidification and otter (*Lutra lutra*) distribution in a British river. *Mammalia* **51**: 81–87.

Mellanby K. (editor) (1988) *Air Pollution, Acid Rain and Environment.* Watt Committee on Energy Report No. 18, Elsevier, London, 129 pp.

Miller H. G. (1984) Deposition–soil–plant interactions. *Phil. Trans. Roy. Soc. London* **305B**: 339–352.

Miller H. G. and J. M. Cooper (1976) Tree growth and climatic cycles in the rain shadow of the Grampian mountains. *Nature* **260**: 697–698.

Mills K. H., S. M. Chalanchuk, L. C. Mohr, I. J. Davies (1987) Responses of fish populations in Lake 223 to 8 years of experimental acidification. *Can. J. Fish. Aquat. Sci.* **44**: 114–125.

Mills K. H. and D. W. Schindler (1986) Biological indicators of lake acidification. *Water Air Soil Pollut.* **30**: 779–789.

Minshall G. W. and J. N. Minshall (1978) Further evidence on the role of chemical factors in determining the distribution of benthic invertebrates in the River Duddon. *Arch. Hydrobiol.* **83**: 324–355.

MONITOR (1985) Report of the National Swedish Environmental Monitoring Programme (PMK), SNV, Solna, Sweden, 207 pp.

Morling G. (1981) Effects of acidification on some lakes in western Sweden. *Vatten* **1**: 25–38.

Muniz I. P. (1981) Acidification and the Norwegian salmon, pp. 66–71 in *Acid Rain and the Atlantic Salmon*, Int. Atlantic Salmon Foundation, Special Publ. **10**.

Muniz I. P. (1984) The effects of acidification on Scandinavian freshwater fisheries. *Phil. Trans. Roy. Soc. London* **305B**: 259–270.

Muniz I. P. and H. Leivestad (1980) Acidification effects on freshwater fish, pp. 84–92 in *Ecological Impact of Acid Precipitation*, ed. D. Drablos and A. Tollan, SNSF Report, Oslo.

Muniz I. P. and L. Walloe (1990) The influence of water quality and catchment characteristics on the survival of fish, pp. 327–339 in *The Surface Water Acidification Programme*, ed. B. J. Mason, publ. Cambridge University Press, Cambridge.

Murdoch P. S. (1988) Chemical budgets and stream chemistry dynamics of a headwater stream in the Catskill Mountains, New York, 1984–85. Report 88-4038, Water Resources Investigations, US Geological Survey.

Newson M. D. (1975) *The Physiography, Deposits and Vegetation at the Plynlimon Catchment*, Institute of Hydrology, Report **30**, Wallingford, UK.

Nilsson J. (editor) (1986) *Critical Loads for Sulphur and Nitrogen.* Nordic Council of Ministers, Copenhagen, **11**, 232 pp.

Nilsson J. and P. Grennfelt (editors) (1988) Critical loads for sulphur and nitrogen. UNECE/Nordic Council workshop report, Skokloster, Sweden, March 1988, Nordic Council of Ministers, Copenhagen, 418 pp.

Nilsson S. I. (1985) Why is lake Gardsjon acid? *Ecol. Bull.* **37**: 311–318.

Nilsson S. I., H. G. Miller, J. G. Miller (1982) Forest growth as a possible cause of soil and water acidification—an examination of concepts. *Oikos* **39**: 40–49.

Nisbet A. F. and T. R. Nisbet (1992) Interactions between rain, vegetation and soils, pp. 135–149 in *Restoring Acid Waters*, ed. G. Howells and T. R. K. Dalziel, Elsevier, London, 421 pp.

Nordic Council of Ministers (1988) *Surface Water Acidification in the ECE Region*, Nordic Council of Ministers, Copenhagen, 156 pp.

Norton S. A., A. Henriksen, B. M. Wathne, A. Veidel (1987) Aluminium dynamics in response to experimental additions of acid to a small Norwegian stream, pp. 249–258 in *Water and Acidification Pathways*, Bolkesjo, Norway.

Novo A., A. Buffoni and M. Tita (1992) Rain and throughfall chemistry in a Norway spruce forest in the Western Prealps. *Environ. Pollut.* **75**: 209–213.

Nyberg P. (1984) Recovery of limed lakes in Sweden. *Phil. Trans. Roy. Soc. London* **305B**: 291–302.

NYDEC (1985) Adirondacks Lakes Survey, 1984, Adirondacks Lakes Survey Corp., NY Dept. Environ. Conservation.

Nygaard G. (1956) Ancient and recent flora of the diatoms and chrysophyceae in Lake Gribsø. Studies on the humic acid Lake Gribsø. *Folia Limnol. Scand.* **8**: 32–94.

Nyholm N. E. I. (1981) Evidence of the involvement of aluminium in the causation of defective formation of egg shells and impaired breeding in wild passerine birds. *Environ. Res.* **26**: 363–331.

Nyholm N. E. I. and H. E. Myrberg (1977) Severe eggshell defects and impaired reproductive capacity in small passerines in Swedish Lappland. *Oikos* **29**: 336–341.

Oden S. (1967) *Dagens Nyheter*, October 1967.

Oden S. (1976) The acidity problem—an outline of concepts. *Water Air Soil Pollut.* **6**: 137–166.

Oden S. and T. Ahl (1972) The long-term changes in pH of lakes and rivers in Sweden, in *Sweden's Case Study*, Min. Foreign Affairs and Min. Agriculture, Stockholm, 13 pp.

Ogden J. G. (1982) Seasonal mass balance of major ions in three small watersheds in a maritime environment. *Water Air Soil Pollut.* **17**: 119–130.

Okland J. (1992) Effect of acid water on freshwater snails: results from a study of 1000 lakes throughout Norway. *Environ. Pollut.* **78**: 127–130.

Olem H. (1990) *Liming Acidic Surface Waters*, Lewis Publishers, MI, 331 pp.

Ormerod S. J. (1985) The diet of the breeding dipper, *Cinclus cinclus*, and their nestlings in the catchment of the river Wye: a preliminary study by faecal analysis. *Ibis* **127**: 316–331.

Ormerod S. J., P. Boole, C. P. McCahon, N. S. Weatherley, D. Pascoe, R. W. Edwards (1987b) Short-term experimental acidification of a Welsh stream: comparing biological effects of hydrogen ions and aluminium. *Freshwat. Biol.* **17**: 341–356.

Ormerod S. J., A. P. Donald, S. J. Brown (1989) The influence of plantation forestry on the pH and aluminium concentrations of upland Welsh streams: a re-examination. *Environ. Pollut.* **62**: 47–62.

Ormerod S. J. and R. W. Edwards (1987) The ordination and classification of macroinvertebrate assemblages in the catchment of the River Wye. *Freshwat. Biol.* **17**: 533–546.

Ormerod S. J. and A. S. Gee (1990) Chemical and ecological evidence on the acidification of Welsh lakes and rivers, pp. 11–25 in *Acid Waters in Wales*, ed. R. W. Edwards *et al.*, Kluwer, Dordrecht, Netherlands.

Ormerod S. J. and S. J. Tyler (1989) Long-term changes in the suitability of Welsh streams for dippers, *Cinclus cinclus*, as a result of acidification and recovery: a modelling study. *Environ. Pollut.* **62**: 171–182.

Ormerod S. J., K. R. Wade, A. G. Gee (1987a) Macrofloral assemblages in upland Welsh streams in relation to their acidity, and their importance to invertebrates. *Freshwat. Biol.* **18**: 545–557.

Ormerod S. J., N. S. Weatherley, A. S. Gee (1990) Modelling the ecological impact of changing acidity in Welsh streams, pp. 279–298 in *Acid Waters in Wales*, ed. R. W. Edwards *et al.*, Kluwer, Dordrecht, Netherlands.

Ormerod S. J., N. S. Weatherley, P. V. Varallo, P. G. Whitehead (1988) Preliminary estimates of the historical and future impact of acidification on the ecology of Welsh streams. *Freshwat. Biol.* **20**: 127–140.

Otto C. and B. S. Svenson (1983) Properties of acid brown water streams in south Sweden. *Arch. Hydrobiol.* **99**: 1–11.

Parker G. G. (1983) Throughfall and stemflow in the forest nutrient cycle. *Adv. Ecol. Res.* **13**: 57–133.

Patrick S., D. Waters, S. Juggins, A. Jenkins (1991) The United Kingdom Acid Waters Monitoring Network: Site descriptions and methodology report, Ensis, London, 63 pp.

Pennington W. (1981) Record of a lake's life in time: the sediments. *Hydrobiologia* **79**: 197–219.

Pennington W. (1984) Long term natural acidification of upland sites in Cumbria: evidence from post-glacial sediments. *Freshwat. Biol. Assocn. Ann. Rep.* **52**: 28–46.

Petersen R. H., P. G. Daye, J. L. Metcalfe (1980) Inhibition of Atlantic salmon (*Salmo salar*) hatching at low pH. *Can. J. Fish. Aquat. Sci.* **37**: 770–774.

Petersen R. C., A. Hargeby, A. Kullberg (1987) The biological importance of humic material in acidified waters. Report 3388, National Swedish Environ. Prot. Board, Solna, Sweden.

Pfeiffer M. H. and P. J. Festa (1980) Acidity status of lakes in the Adirondack region of New York in relation to fish resources. New York Dept. Environ. Conservation FW-P 168 (10/80), 36 pp.

Placet M. (1990) Emissions involved in acidic deposition processes, in *Acidic Deposition, State of Science and Technology*, NAPAP Report 1, Washington, DC.

Potts W. T. W. and G. Fryer (1979) The effects of pH and salt content on sodium balance in *Daphnia magna* and *Acantholeberis curvirostris* (Crustacea: Cladocera). *J. Comp. Physiol.* **129**: 289–294.

Potts W. T. W. and P. G. McWilliams (1989) The effects of hydrogen and aluminium ions on fish gills, pp. 201–220 in *Acid Toxicity and Aquatic Animals*, ed. R. Morris *et al.*, Cambridge University Press, Cambridge.

Potts W. T. W., C. Talbot, F. B. Eddy, M. Williams (1990) Sodium balance in adult Atlantic salmon (*Salmo salar* L.) during migration from seawater to acid freshwater, pp. 369–382 in *The Surface Waters Acidification Programme*, ed. B. J. Mason, Cambridge University Press, Cambridge.

Pough P. F. (1976) Acid precipitation and embryonic mortality of spotted salamanders, *Ambystoma maculatum*. *Science* **192**: 68–70.

Prigg R. F. (1983) Juvenile salmonid populations and biological quality of upland streams in Cumbria, with particular reference to low pH effects. NWWA Rivers Division Report, BN11-1-83, 40 pp.

Psenner R. and R. Schmidt (1992) Climate driven control of remote alpine lakes and effects of acid deposition. *Nature* **356**: 781–783.

Raddum G. G. and A. Fjellheim (1984) Acidification and early warning organisms in fresh water in western Norway. *Verh. Internat. Verein. Limnol.* **22**: 1973–1980.

Raddum G. G. and A. Fjellheim (1987) Effects of pH and aluminium on mortality, drift, and moulting of the mayfly, *Baetis rhodani*. *Ann. Soc. Roy. Belg.* **117**: 77–87.

Rahel F. H. (1982) Fish assemblages in Wisconsin bog lakes. PhD Thesis, University of Wisconsin, Madison, WI.

Rao S. S., A. A. Jurkovic, J. O. Nriagu (1984a) Bacterial activity in sediments of lakes receiving acid precipitation. *Environ. Pollut.* **36**: 195–205.

Rao S. S., D. Paolini, G. G. Leppard (1984b) Effects of low pH stress on the morphology and activity of bacteria from lakes receiving acid precipitation. *Hydrobiologia* **114**: 115–121.

Rao V. N. R. and S. K. Subramian (1982) Metal toxicity tests on growth of some diatoms. *Acta Botan. India* **10**: 274–281.

Rask M. (1992) Changes in the density, population structure, growth and reproduction of perch, *Perca fluviatilis* L., in two acidic forest sites in southern Finland. *Environ. Pollut.* **78**: 121–125.

Raven P. J. (1985) The use of aquatic macrophytes to assess water quality changes in some Galloway lochs; an exploratory study. Palaeoecology Research Unit, Working Paper **9**, University College London, 76 pp.

Raven P. J. (1988) Occurrence of *Sphagnum* moss in the sublittoral of several small oligotrophic lakes in Galloway, southwest Scotland. *Aquat. Bot.* **30**: 223–230.

Raven P. J. (1989) Short-term changes in the aquatic macrophyte flora of Loch Fleet, SW Scotland, following catchment liming, with particular reference to sublittoral *Sphagnum*. Palaeological Research Unit, University College London, Research Paper No. 39, 42 pp.

Reader J. P., T. R. K. Dalziel, R. Morris (1988) Growth, mineral uptake and skeletal calcium deposition in brown trout, *Salmo trutta* L., yolk sac fry exposed to aluminium and manganese in soft acid water. *J. Fish. Biol.* **32**: 607–624.

Reeve R. and I. F. Fergus (1982) Black and white waters and their possible relationship to podzolization process. *Austral. J. Soil Res.* **21**: 59–66.

Renberg I. and R. W. Battarbee (1990) The SWAP palaeolimnology programme: a synthesis. *Phil. Trans. Roy. Soc. London* **327B**: 281–299.

Renberg I. and T. Hellberg (1982) The pH history of lakes in southwestern Sweden, as calculated from the fossil diatom flora of the sediments. *Ambio* **11**: 30–33.

Reuss J. G. (1977) Chemical and biological relationships relevant to the effect of acid rainfall on the soil–plant system. *Water Air Soil Pollut.* **7**: 461–478.

Reuss J. G., B. J. Cosby, R. F. Wright (1987) Chemical processes governing soil and water acidification. *Nature* **329**: 27–32.

Reuss J. G. and D. W. Johnson (1986a) Effect of soil processes on the acidification of water by acid deposition. *J. Environ. Qual.* **14**: 26–41.

Reuss J. G. and D. W. Johnson (1986b) *Acid Deposition and the Acidification of Soils and Waters*, Springer-Verlag, Berlin.

Reynolds B., J. Cape, M. Hornung, S. Hughes, P. A. Stevens (1988) Impact of afforestation on the soil solution chemistry of stagnopodzols in mid-Wales. *Water Air Soil Pollut.* **38**: 29–39.

Reynolds B., B. A. Emmett, C. Woods (1992) Variations in streamwater nitrate concentrations and nitrogen budgets over 10 years in a headwater catchment in mid-Wales. *J. Hydrol.* **136**: 155–175.

Reynolds B., M. Hornung, B. A. Emmett, S. J. Brown (1993) Amelioration of streamwater acidity by catchment liming—response of podzolic soils to pasture improvement. *Chemistry and Ecology* **8**, 233–248.

Richard Y. (1982) Relationships between the acidity level of 158 lakes in the Parc des Laurentides and the evolution of their fishing statistics over the last nine years. Service de la Qualité des Eaux, Min. Environ. Québec, Report No. Pa-5, 115 pp.

Richter D. D., D. W. Johnson, D. E. Todd (1983) Atmospheric sulfur deposition, neutralization, and ion leaching in two deciduous forest ecosystems. *J. Environ. Qual.* **12**: 263–270.

Roberts T. M. (1984) Long term effects of sulphur dioxide on crops: an analysis of dose–response relations. *Phil. Trans. Roy. Soc. London* **305B**: 299–316.

Roberts T. M. (1987) Effects of air pollutants on agriculture and forestry. *CEGB Research* **20**: 39–52.

Roberts T. M., N. M. Darrall, P. Lane (1983) Effects of gaseous air pollutants on agriculture and forestry in the UK. *Adv. Appl. Biol.* **9**: 2–142.

Roberts T. M., R. A. Skeffington, L. W. Blank (1989) Causes of Type 1 spruce decline in Europe. *Forestry* **62**: 179–222.

Rochelle B. P., M. R. Church, M. B. David (1987) Sulfur retention at intensively studied watersheds in the US and Canada. *Water Air Soil Pollut.* **33**: 73–83.

Rodhe W. (1949) The ionic composition of lake waters. *Verh. Internat. Verein. Limnol.* **10**: 377–386.

Roff J. C. and R. E. Kwiatkowski (1977) Zooplankton and zoobenthos communities of selected northern Ontario lakes of different acidities. *Can. J. Zool.* **55**: 899–911.

Rosenberg A. (1990) Discussion, pp. 341–342 in *The Surface Water Acidification Programme*, ed. B. J. Mason, Cambridge University Press, Cambridge.

Rosenqvist I. Th. (1978) Alternative sources of river acidification in Norway. *Sci. Tot. Environ.* **10**: 39–49.

Rosseland B. O. (1986) Ecological effects of acidification on tertiary consumers: Fish population responses. *Water Air Soil Pollut.* **30**: 451–460.

Rosseland B. O., I. A. Blakar, A. Bulger, F. Kroglund, A. Kvellstad, E. Lydersen, D. H. Oughton, B. Salbu, M. Stuarnes, R. Voigt (1992) The mixing zone between limed and acidic river waters: complex aluminium chemistry and extreme toxicity for salmonids. *Environ. Pollut.* **78**: 3–8.

Rosseland B. O. and A. Hindar (1988) Liming of lakes, rivers, and catchments in Norway. *Water Air Soil Pollut.* **41**: 165–188.

Rosseland B. O., L. Lien, F. Kroglund, K. Sadler, T. R. K. Dalziel (1990) Strains of brown trout (*Salmo trutta* L.): stockings and test fishing 1988 and 1989. Field and laboratory toxicity experiments. NIVA Report 0-87178.

Rosseland B. O., I. Sevaldrud, D. Svastalog, I. P. Muniz (1980) Studies in freshwater fish population structure, growth and food selection, pp. 336–337 in *Ecological Impact of Acid Precipitation*, ed. D. Drablos and A. Tollan, SNSF Report, Oslo, Norway.

Rosseland B. O., O. K. Skogheim, I. V. Sevaldrud (1986) Acid deposition and effects in Nordic Europe. Damage to fish populations in Scandinavia continues apace. *Water Air Soil Pollut.* **30**: 65–74.

Rossi H. (1984) Report of the UK Select Committee on Environmental Pollution: Acid Rain, HMSO, London.

Royal Ministry for Foreign Affairs and the Royal Ministry of Agriculture (1971) *Air Pollution across National Boundaries. The Impact on the Environment of Sulfur in Air and Precipitation*, Stockholm, 96 pp.

Rudd J. W. M., C. A. Kelly, A. Furutani (1986a) The role of sulfate reduction in long term accumulation of organic and inorganic sulfur in lake sediments. *Limnol. Oceanogr.* **31**: 1281–1291.

Rudd J. W. M., C. A. Kelly, V. St Louis, R. H. Hesslein, A. Furutani, H. Holoka (1986b) Microbial consumption of nitric and sulfuric acids in acidified north temperate lakes. *Limnol. Oceanogr.* **31**: 1515–1517.

Rudd J. W. M., C. A. Kelly, D. W. Schindler, M. A. Turner (1990) A comparison of the acidification efficiencies of nitric and sulfuric acids by two whole-lake acidification experiments. *Limnol. Oceanogr.* **35**: 663–679.

Runn P., N. Johansson, G. Milbrink (1977) Some effects of low pH on the hatchability of eggs of perch, *Perca fluviatilis* L. *Zoon* **5**: 115–125.

Sadler K. and S. Lynam (1986) Some effects of low pH and calcium on the growth and tissue mineral content of yearling brown trout, *Salmo trutta. J. Fish. Biol.* **29**: 313–324.

Sadler K. and S. Lynam (1987) The effects on the growth of brown trout from exposure to aluminium at different pH levels. *J. Fish. Biol.* **31**: 209–219.

Sadler K. and S. Lynam (1988) The influence of calcium on aluminium-induced changes in the growth rate and mortality of brown trout, *Salmo trutta* L. *J. Fish. Biol.* **33**: 171–179.

Sanden P., A. Grimvall, U. Lohm (1987) Acidification trends in Sweden. *Water Air Soil Pollut.* **36**: 259–270.

Scanlon P. (1990) Effects of acidification on wild mammals and wildfowl, pp. 151–156 in *Acid Deposition: State of Science and Technology*, NAPAP Rep. 13, J. P. Baker *et al.*, 1990.

Schindler D. W. (1986) The significance of in-lake production of alkalinity. *Water Air Soil Pollut.* **30**: 931–944.

Schindler D. W. (1987) Detecting ecosystem responses to anthropogenic stress. *Can. J. Fish. Aquat. Sci.* **44**: 6–25.

Schindler D. W., T. M. Frost, K. H. Mills, P. S. S. Chang, I. J. Davies, L. Findlay, D. F. Malley, J. A. Shearer, M. A. Turner, P. J. Garrison, C. J. Watras, K. Webster, J. M. Gunn, P. L. Brezonik, W. A. Swenson (1991) Comparisons between experimentally and atmospherically-acidified lakes during stress and recovery, pp. 193–226 in *Acidic Deposition: Its Nature and Impacts*, ed. F. T. Last and R. Watling, Roy. Soc. Edinburgh.

Schindler D. W., K. H. Mills, D. F. Malley, D. L. Findlay, J. A. Shearer, I. J. Davies, M. A. Turner, G. A. Linsey, D. R. F. Cruikshank (1985) Long-range ecosystem stress: the effects of years of experimental acidification of a small lake. *Science* **228**: 1395–1401.

Schindler D. W. and M. A. Turner (1982) Biological, chemical and physical responses to experimental acidification. *Water Air Soil Pollut.* **18**: 259–271.

Schindler D. W., M. A. Turner, M. P. Stainton, G. A. Lindsey (1986) Natural sources of acid neutralising capacity in low alkalinity lakes of the Precambrian Shield. *Science* **232**: 844–847.

Schindler D. W., R. Wagemann, R. B. Cook, T. Ruszczynski, J. Propowich (1980) Experimental acidification of Lake 223, Experimental Lakes Area: Background data and the first three years of experimental acidification. *Science* **228**: 1395–1401.

Schmitt G. (1988) Measurements of the chemical composition in cloud and fogwater, pp. 403–419 in *Cloud Deposition at High Altitude Sites*, ed. M. H. Unsworth and D. Fowler, Kluwer, Dordrecht, Netherlands.

Schmuul R. (1976) Physical and chemical investigations of 1704 lakes in the northern part of the province of Alvsburg in 1973. *Inform. Drottningholm* **4**, 42 pp.

Schnoor J. L. and W. Stumm (1986) The role of chemical weathering in the neutralization of acid deposition. *Schweiz. Z. Hydrol.* **48**: 171–195.

Schofield C. R. (1976) Acid precipitation: effects on fish. *Ambio* **5**: 228–230.

Schulze E.-D. and P. H. Freer-Smith (1991) An evaluation of forest decline based on field observations focused on Norway spruce, *Picea abies. Proc. Roy. Soc. Edinburgh* **97B**: 155–168.

Seip H. A. (1980) Acidification of freshwater—sources and mechanisms, pp. 358–365 in *Ecological Impact of Acid Precipitation*, ed. D. Drablos and A. Tollan, SNSF, Sandefjord, Norway.

Seip H. M., S. Andersen, B. Halsvik (1980) Snowmelt studied in a minicatchment with neutralized snow. SNSF Report IR 65/80, Oslo, Norway, 20 pp.

Seip H. M., S. Andersen, A. Henriksen (1990a) Geochemical control of aluminium concentrations in acidified surface waters. *J. Hydrol.* **116**: 299–305.

Seip H. M., I. A. Blakar, N. Christophersen, H. G. Grip, R. D. Vogt (1990b) Hydrochemical studies in Scandinavian catchments, pp. 19–29 in *The Surface Waters Acidification Programme*, ed. B. J. Mason, Cambridge University Press, Cambridge.

Seip H. M., L. Muller, A. Naas (1984) Aluminium speciation: a comparison of two spectrometric methods and observed concentrations in some acidic aquatic ecosystems in southern Norway. *Water Air Soil Pollut.* **23**: 81–95.

Sevaldrud I. V., I. P. Muniz, S. Kalvenes (1980) Loss of fish populations in southern Norway: Dynamics and magnitude of the problem, pp. 350–351 in *Ecological Impact of Acid Precipitation*, ed. D. Drablos and A. Tollan, SNSF Report, Oslo, Norway.

Sevaldrud I. V. and O. H. Skogheim (1986) Changes in fish populations in southernmost Norway during the last decade. *Water Air Soil Pollut.* **30**: 381–386.

Shore R. F. and S. Mackenzie (1993) The effects of catchment liming on shrews, *Sorex* spp. *Biol. Conserv.* **64**: 101–111.

Siegel B. Z., M. Nachbar-Hapai, S. M. Siegel (1990) The contribution of sulfate to rainfall pH around Kilauea Volcano, Hawaii. *Water Air Soil Pollut.* **52**: 227–235.

Singh B. R. (1984) Sulphate sorption by forest soils: I. Sulphate sorption isotherms and comparison of different absorption equations describing sulphate absorption. *Soil Sci.* **138**: 189–197.

Sioli H. (1975) Tropical river: the Amazon, pp. 461–488 in *River Ecology*, ed. B. A. Whitton, Blackwells, Oxford.

Skeffington, R. A. (1983) Soil properties under three species of tree in southern England in relation to acid deposition in throughfall, pp. 219–231 in *Effects of Accumulation of Air Pollutants in Forest Ecosystems* ed. B. Ulrich and J. Pankrath, Reidel, Dordrecht, Netherlands.

Skeffington R. A. (1985) The transport of acidity through ecosystems, pp. 139–154 in *Pollutant Transport and Fate in Ecosystems*, ed. P. J. Coughtrey, M. H. Martin, M. Unsworth, Blackwell, Oxford.

Skeffington, R. A. (1987) Soil and its response to acid deposition. *CEGB Research* **20**: 16–29.

Skeffington R. A. (1992) Soil/water acidification and the potential for reversibility, pp. 23–37 in *Restoring Acid Waters*, ed. G. Howells and T. R. K. Dalziel, Elsevier, London.

Skeffington R. A. and D. J. A. Brown (1992) Timescales of recovery from acidification: implications of current knowledge for aquatic organisms. *Environ. Pollut.* **77**: 227–234.

Skeffington R. A. and K. A. Brown (1986) The effect of five years' acid treatment on leaching, soil chemistry and weathering of a homo-ferric podzol. *Water Air Soil Pollut.* **31**: 891–900.

Skeffington R. A. and E. J. Wilson (1988) Excess nitrogen deposition: issues for consideration. *Environ. Pollut.* **54**: 159–184.

Skiba U. and M. S. Cresser (1991) Seasonal changes in soil atmospheric CO_2 concentrations in two upland catchments and associated changes in river water chemistry. *Chemistry and Ecology* **5**: 217–225.

Skiba U., M. S. Cresser, R. G. Derwent, D. W. Futty (1989) Peat acidification in Scotland. *Nature.* **337**: 68–69.

Skogheim O. K., B. O. Rosseland, F. Kroglund, H. Hagenland (1986) Addition of NaOH, limestone slurry and fine-grained limestones to acidified lake water and the effects on smolts of Atlantic salmon (*Salmo salar* L.) *Water Air Soil Pollut.* **30**: 587–592.

Smith R. A. (1852) On the air and rain of Manchester. *Mem. Lit. Phil. Soc., Manchester* (ser. 2) **10**: 207–217.

Smith R. A. (1872) *Air and Rain—the Beginnings of Chemical Climatology*, Longmans, Green and Co., London.

SNV (Swedish Environment Protection Board) (1983) *Ecological Effects of Acid Deposition*, Report and background papers for 1982 Stockholm Conference on Acidification of the Environment, SNV 1636, 340 pp.

Soderlund R. (1977) NO_x and ammonia emissions—a mass balance for the atmosphere over NW Europe. *Ambio* **16**: 118–122.

Sollins P., C. C. Grijer, F. M. McCorison, K. Cromack, R. Fogel, R. L. Frederikson (1980) The internal element cycle of an old growth Douglas fir system in western Oregon. *Ecol. Monographs* **50**: 261–285.

Sorenson S. L. P. (1909) Über die Messung und die Bedeutung der Wasserstoff-konzentration bei enzymatischen Prozessen. *Biochem. Zeit.* **21**: 131–139.

Stallard R. F. and J. M. Edmond (1983) Geochemistry of the Amazon: 2. The influence of geology and weathering environment on the dissolved load. *J. Geophys. Res.* **88(C14)**: 9671–9688.

Stenson J. A. E. (1985) Biotic structures and relations in the acidified Lake Gardsjon system—a synthesis. *Ecol. Bull.* **37**: 319–139.

Stewart B. R., K. Paterson, T. R. K. Dalziel, M. V. Proctor (1992) Deposition input considerations, pp. 99–112 in *Restoring Acid Waters*, ed. G. Howells and T. R. K. Dalziel, Elsevier, London.

Stewart J. E. (editor) (1989) Report of the Study Group on toxicological mechanisms involved in the impact of acid rain and its effects on salmon. ICES Report CM 1988/M:4, Copenhagen, 67 pp.

Stewart J. E., J. Brown, K. Friedland, L. P. Hansen, T. Hesthagen, G. Howells, G.

Lacroix, L. Marshall, A. L. Meister, W. Watt (1988) Report of the Acid Rain Study Group, ICES Report CM 1988/M:5, 53 pp.

Stokoe R. (1983) Aquatic macrophytes in the lakes and tarns of Cumbria. *Occ. Publ. Freshwat. Biol. Assoc. UK*, 60 pp.

Stoner J. H. and A. S. Gee (1985) The effects of forestry on water quality and fish in Welsh rivers and lakes. *J. Inst. Water Eng. Sci.* **39**: 27–45.

Strom K. M. (1925) pH values in Norwegian mountains. *Nyt. Magas. Naturen*, pp. 237–244.

Stumm W. and J. J. Morgan (1981) *Aquatic Chemistry*, 2nd edition, John Wiley and Sons, New York and Chichester, 780 pp.

Sullivan T. J. (1990) Historical changes in surface water acid base chemistry in response to acidic deposition. *Acid Deposition: State of Science and Technology*, NAPAP Report 11, Washington, DC.

Sullivan T. J., N. Christophersen, I. P. Muniz, H. M. Seip, P. D. Sullivan (1986) Aqueous aluminium response to episodic increases in discharge. *Nature* **323**: 324–327.

Sutcliffe D. W. (1983) Acid precipitation and its effects on aquatic systems in the English Lake District. *Freshwat. Biol. Assocn. Ann. Rept.* **51**: 30–62.

Sutcliffe D. W. and T. R. Carrick (1973) Studies on mountain streams in the English Lake District. I: pH, calcium and the distribution of invertebrates in the River Duddon. *Freshwat. Biol.* **3**: 437–462.

Sutcliffe D. W. and T. R. Carrick (1988) Alkalinity and pH of tarns and streams in the English Lake District. *Freshwat. Biol.* **19**: 179–189.

Sutcliffe D. W. and A. G. Hildrew (1989) Invertebrate communities in acid streams, pp. 13–29 in *Acid Toxicity and Aquatic Animals* ed. R. Morris *et al.*, Cambridge University Press, Cambridge.

Sutcliffe D. W., T. R. Carrick, J. Heron, E. Rigg, J. F. Talling, C. Woof, J. W. G. Lund (1982) Long-term and seasonal changes in the chemical composition of precipitation and surface waters of lakes and tarns in the English Lake District. *Freshwat. Biol.* **12**: 451–506.

Sverdrup H., P. Warfvinge, T. Frogner, A. O. Haoya, M. Johansson, B. Andersen (1992) Critical loads for forest soils in the Nordic countries. *Ambio* **21**: 348–355.

Swedish Min. Agric. (1982) *Acidification Today and Tomorrow*, Stockholm, 231 pp.

Taylor G. E. and L. F. Pitelka (1992) Genetic diversity of plant populations and the role of air pollution, pp. 111–130 in *Air Pollution Effects on Biodiversity*, ed. J. R. Barker and D. T. Tingey, Van Nostrand Reinhold, New York.

Thompson M. E. (1987) Comparison of excess sulfate yields and median pH values of rivers in Nova Scotia and Newfoundland. *Water Air Soil Pollut.* **35**: 377–341.

Thornton K. W., D. Marmorek, P. F. Ryan (1990) Methods for projecting future changes in surface water acid–base chemistry. *Acid Deposition: State of Science and Technology*, NAPAP Report 14, Washington, DC.

Tipping E. (1989a) A model of surface water acidification in Cumbria and its uses in long-term research. *Freshwat. Biol.* **23**: 7–23.

Tipping E. (1989b) Acid sensitive waters of the English Lake District: a steady state model of streamwater chemistry in the upper Duddon catchment. *Environ. Pollut.*

60: 181–208.

Tipping E. (1990) Aluminium chemistry in acid environments, pp. 255–260 in *The Surface Waters Acidification Programme*, ed. B. J. Mason, Cambridge University Press, Cambridge.

Tipping E. and J. Hopwood (1988) Estimating streamwater concentrations of aluminium released from streambeds during acid episodes. *Environ. Technol. Lett.* **9**: 703–712.

Tipping E., C. Woof, P. B. Walters, M. Ohnstad (1988) Aluminium speciation in acidic natural waters: testing of a model for Al–humic complexation. *Water Res.* **22**: 321–326.

Townsend C. R. and A. G. Hildrew (1984) Longitudinal pattern in detritivore communities in acid streams: a consideration of alternative hypotheses. *Verh. Internat. Verein. Limnol.* **22**: 1953–1958.

Townsend C. R., A. G. Hildrew, J. E. Francis (1983) Community structure in some southern English streams: the influence of physico-chemical factors. *Freshwat. Biol.* **13**: 521–544.

Townsend G. S., K. H. Bishop, B. W. Bache (1990) Aluminium speciation during episodes, pp. 275–278, in *The Surface Waters Acidification Programme*, ed. B. J. Mason, Cambridge University Press, Cambridge.

Traaen T. S. (1978) Bakterieplankton i innsjoer, SNSF Report TN/41/78, 16 pp.

Traaen T. S. (1980) Effects of acidity on decomposition of organic matter in aquatic environments, pp. 340–341 in *Ecological Impacts of Acid Precipitation*, ed. D. Drablos and A. Tollan, SNSF Report, Oslo, Norway.

Tranter M., P. W. Abrahams, I. L. Blackwood, P. Brimblecombe, T. D. Davies (1988) The impact of a single black snowfall on streamwater chemistry in the Scottish Highlands. *Nature* **332**: 826–829.

Tranter M., T. D. Davies, P. Brimblecombe, C. E. Vincent (1987) The composition of acidic meltwaters during snowmelt in the Scottish Highlands. *Water Air Soil Pollut.* **36**: 75–90.

Trippel E. A. and H. H. Harvey (1987) Reproductive responses of five white sucker (*Catastomus comersoni*) populations in relation to lake acidity. *Can. J. Fish. Aquat. Sci.* **44**: 1018–1023.

Turk J. (1985) Natural variance in pH as a complication in detecting acidification in lakes. *Water Air Soil Pollut.* **37**: 171–176.

Turner L. J. and J. R. Kramer (1992) Irreversibility of sulphate absorption on goethite and haematite. *Wat. Air Soil Pollut.* **63**: 23–32.

Turner M. A., E. T. Howell, M. Summerby, R. H. Hesslein, D. L. Findlay, M. B. Jackson (1991) Changes in epilithon and epiphyton associated with experimental acidification of a lake to pH 5. *Limnol. Oceanogr.* **36**: 1390–1405.

Turnpenny A. W. H. (1989) Field studies on fisheries in acid waters in the UK, pp. 45–65 in *Acid Toxicity and Aquatic Animals*, ed. R. Morris *et al.*, Cambridge University Press, Cambridge.

Turnpenny A. W. H., C. H. Dempsey, M. H. Davies, J. M. Fleming (1988) Factors limiting fish populations in the Loch Fleet system, an acidic drainage system in SW Scotland. *J. Fish. Biol.* **32**: 415–434.

Turnpenny A. W. H., K. Sadler, R. J. Aston, A. G. P. Milner, S. Lynam (1987) The fish populations in some streams in Wales and northern England in relation to acidity and associated factors. *J. Fish. Biol.* **41**: 415–434.

Tyler-Jones R., R. C. Beattie, R. J. Aston (1986) The effects of acid water and aluminium on the embryonic development of the common frog, *Rana temporaria*. *J. Zool. London* **219**: 355–372.

Ulrich B. (1983a) Soil acidity and its relation to acid deposition, pp. 127–145 in *Effects of Accumulation of Air Pollutants in Forest Ecosystems*, ed. B. Ulrich and J. Pankrath, Reidel, Dordrecht, Netherlands.

Ulrich B. (1983b) A concept of forest ecosystem stability and of acid deposition as a driving force for destabilisation, pp. 1–29 in *Effects of Accumulation of Air Pollutants in Forest Ecosystems*, ed. B. Ulrich and J. Pankrath, Reidel, Dordrecht, Netherlands.

Ulrich B., R. Mayer, P. K. Khanna (1979) Deposition von Luftvereinigungen und ihre Auswirkungen in Waldoekosystem im Solling. *Forstl. Versuchsanstalt* **58**: 1–291.

United Nations (UN) (1985) Protocol on the reduction of sulphur emissions or their transboundary fluxes by at least 30%. UN Document ECE/EB.AIR/7 Annex 1.

UNECE (United Nations Economic Commission for Europe) (1982a) *Effects of Sulphur Compounds on Materials, including Historical and Cultural Monuments*, ENV/IEB/R3, United Nations, Geneva.

UNECE (1982b) *Effects of Sulphur Compounds and Other Related Air Pollutants on Health*, ENV/IEB/R14, United Nations, Geneva.

UNECE (1984) *Air Borne Sulphur Pollution: Effects and Control*, ECE/EB.AIR, Report 1, United Nations, Geneva, 265 pp.

UNECE (1985) *Air Pollution Across Boundaries*, ENV/EB.AIR/5, United Nations, Geneva, 156 pp.

UNECE (1986) *Transboundary Air Pollution: Effects and Control*, ECE/EB.AIR/8, United Nations, Geneva, 77 pp.

Urch U. A. and J. L. Hedrick (1981) Isolation and characterization of the hatching enzyme from the amphibian, *Xenopus laevis. Arch. Biochem. Biophys.* **206**: 424–431.

van Breemen N. (1973) Dissolved aluminium in acid sulphate and in acid mine waters. *Soil Sci. Amer. Proc.* **37**: 694–697.

van Breemen N. (1991) Soil acidification and alkalinization, in *Soil Acidity*, ed. B. Ulrich and M. E. Sumner, Springer-Verlag, Berlin.

van Breemen N., C. T. Driscoll, J. Mulder (1984) Acidic deposition and internal proton sources in acidification of soils and waters. *Nature* **307**: 599–604.

van Dam H., B. van Geel, A. van der Wijk, J. F. M. Geelen, R. van der Heijden, M. D. Dickman (1988) Palaeolimnological and documented evidence for alkalization and acidification of two moorland pools (The Netherlands). *Rev. Palaeobot. Pakynol.* **55**: 273–316.

Vannote R. L., G. W. Minshall, K. W. Cummins, J. R. Sedell, C. E. Cushing (1980) The river continuum concept. *Can. J. Fish. Aquat. Sci.* **37**: 130–137.

Verry E. S. and K. Nodop (1992) Hydrogen behaviour and ion concentrations in precipitation of Europe and the United States. *Environ. Pollut.* **75**: 129–136.

Vuorinen P. J., M. Rask, M. Vuorinen, J. Raiteneimi, S. Peuranen (1992) Acid/aluminium sensitivity of the pike (*Esox lucius*), whitefish (*Coregonus*) and roach (*Rutilus rutilus*); comparison of field and laboratory data. EIFAC Symp., Lugano, Switzerland, May 1992 (in press).

Vuorinen P. J., M. Vuorinen, S. Peuronen, M. Rask, A. Laapalainen, J. Raitaneimi (1992) Reproductive status, blood chemistry, gill histology and growth of perch (*Perca fluviatilis*) in three acidic lakes. *Environ. Pollut.* **78**: 19–27.

Wales D. L. and G. L. Beggs (1986) Fish species distribution in relation to lake acidity in Ontario. *Water Air Soil Pollut.* **30**: 601–609.

Warfvinge P., M. Holmberg, M. Posch, R. F. Wright (1992) The use of dynamic models to set target loads. *Ambio* **21**: 369–376.

Warfvinge P. and H. Sverdrup (1988) Soil liming as a measure to mitigate acid runoff. *Water Resources Res.* **24**: 701–712.

Warfvinge P. and H. Sverdrup (1992) Calculating critical loads of acid deposition with PROFILE—a steady state soil chemistry model. *Water Air Soil Pollut.* **63**: 119–142.

Warren S. C., G. C. Alexander, B. W. Bache, B. W. Battarbee, D. H. Crawshaw, W. H. Edmunds, H. J. Egglishaw, A. S. Gee, R. Harriman, A. Hildrew, G. Hornung, D. T. E. Hunt, C. Neal, S. J. Ormerod, K. B. Pugh, D. E. Wells, P. G. Whitehead, R. B. Wilson, D. C. Watson (1988) Acidity in United Kingdom Fresh Waters, Report to Dept. Environ., HMSO, London, 61 pp.

Warren S. C., B. W. Bache, W. M. Edmunds, H. J. Egglishaw, A. S. Gee, M. Hornung, G. D. Howells, C. Jordan, J. B. Leeming, P. S. Maitland, K. B. Pugh, D. W. Sutcliffe, D. E. Wells, J. N. Cape, D. T. E. Hunt, R. B. Wilson, D. C. Watson (1986) Acidity in United Kingdom Fresh Waters, Interim Report to Dept. Environ., HMSO, London, 46 pp.

Watson J. (1899) *The English Lake District Fisheries*, Lawrence and Bullen, London.

Watt W. D. (1987) A summary of the impact of acid rain on Atlantic salmon (*Salmo salar*) in Canada. *Water Air Soil Pollut.* **35**: 27–35.

Watt W. D., C. D. Scott, S. Ray (1979) Acidification and other chemical changes in Halifax county lakes after 21 years. *Limnol. Oceanogr.* **24**: 1154–1161.

Watt W. D., C. D. Scott, W. J. White (1983) Evidence of acidification in some Nova Scotian rivers and its impact on Atlantic salmon, *Salmo salar. Can. J. Fish. Aquat. Sci.* **40**: 462–473.

Webster K. E., T. M. Frost, C. J. Watras, W. A. Swenson, M. Gonzales, P. J. Garrison (1992) Complex biological responses to the experimental acidification of Little Rock Lake, Wisconsin, USA. *Environ. Pollut.* **78SE**: 73–78.

Weider R. K., K. P. Heston, E. M. O'Hara, G. E. Lang, A. E. Whitehouse, J. Hett (1988) Aluminium retention in a man-made *Sphagnum* wetland. *Water Air Soil Pollut.* **37**: 177–191.

Wellburn A. R. (1988) *Air Pollution and Acid Rain*. Longmans, Harlow, UK, 274 pp.

Wellburn A. R. and N. M. Darrall (1990) Physiological effects of the direct impact of gaseous pollutants upon foliage. Abstract p. 132 in *Acidic Deposition: its Nature and Impacts*, Int. Conf. Glasgow, September, 1990.

Welsh W. T. and J. C. Burns (1987) The Loch Dee project; runoff and surface water quality in an area subject to acid precipitation and afforestation in southwest Scotland. *Trans. Roy. Soc. Edinburgh* **78B**: 249–260.

Wetzel R. G., E. S. Brammer, K. Lindstrom, C. Forsberg (1985) Photosynthesis of submerged macrophytes in acidified lakes. II. Carbon limitation and utilization of benthic CO_2 sources. *Aquat. Bot.* **22**: 107–120.

Wheater H. S., F. M. Kleissen, M. B. Beck, S. Tuck, A. Jenkins, R. Harriman (1990) Modelling the short-term flow and chemical response in the Allt a'Mharcaidh catchment, pp. 455–466 in *The Surface Waters Acidification Programme*, ed. B. J. Mason, Cambridge University Press, Cambridge.

Whitehead P. G. (1989) Future trends in acidification, pp. 114–121 in *Acidification in Scotland*, Scottish Develop. Dept., Edinburgh.

Wiginton P. J., T. D. Davies, M. Tranter, K. N. Eshleman (1990) Episodic acidification of surface waters due to acidic deposition. *Acid Deposition: State of Science and Technology*, NAPAP Report 12, Washington, DC, 196 pp.

Wilbur H. M. (1984) Complex life-cycles and community organization in amphibians, in *A New Ecology: Novel Approaches to Interactive Systems*, ed. P. W. Price *et al.*, John Wiley and Sons, New York.

Willoughby L. G. and R. G. Mappin (1988) The distribution of *Ephemerella ignita* (Ephemeroptera) in streams: the role of pH and food resources. *Freshwat. Biol.* **19**: 145–155.

Wissmar R. C., A. H. Devol, A. E. Nevissi, J. R. Sedell (1982a) Chemical changes of lakes within the Mount St. Helens blast zone. *Science* **216**: 175–178.

Wissmar R. C., A. H. Devol, J. T. Staley, J. R. Sedell (1982b) Biological responses of lakes in the Mount St. Helens blast zone. *Science* **216**: 178–181.

Wolfenden J., P. A. Wookey, P. W. Lucas, T. A. Mansfield (1992) Action of pollutants, individually and in combination, pp. 72–92 in *Air Pollution Effects on Biodiversity*, ed. J. R. Barker and D. T. Tingey, Van Nostrand Reinhold, New York.

Wood C. M. (1989) The physiological problems of fish in acid waters, pp. 125–152 in *Acid Toxicity and Aquatic Animals*, ed. R. Morris *et al.*, Cambridge University Press, Cambridge.

Wood C. M. and D. G. McDonald (1987) The physiology of acid/aluminium stress in trout. *Ann. Soc. Roy. Zool. Belg.* **117**: 399–410.

Wood C. M., R. C. Playle, B. P. Simons, G. G. Goss, D. G. McDonald (1988) Blood gases, acid–base status, ions and haematology in adult brook trout (*Salvelinus fontinalis*) under acid/aluminium exposure. *Can. J. Fish. Aquat. Sci.* **45**: 1575–1586.

Wren C. D. and G. L. Stephenson (1991) The effect of acidification on the accumulation and toxicity of metals to freshwater invertebrates. *Environ. Pollut.* **71**: 205–241.

Wright R. F. (1977) Historical changes in the pH of 128 lakes in southern Norway and 130 lakes in southern Sweden over the period 1923–1976. SNSF Report TN 34/77, 71 pp.

Wright R. F. and M. Hauhs (1991) Reversibility of acidification: soils and surface waters. *Proc. Roy. Soc. Edinburgh* **97B**: 169–191.

Wright R. F. and A. Henriksen (1978) Chemistry of small Norwegian lakes, with special reference to acidification. *Limnol. Oceanogr.* **23**: 487–498.

Wright R. F. and A. Henriksen (1979) Sulfur: acidification of freshwaters, pp. 277–301, *in Sulphur Emissions and the Environment*, Soc. Chem. Industry, London, 8–10 May 1979.

Wright R. F. and A. Henriksen (1983) Restoration of Norwegian lakes by reduction of sulphur deposition. *Nature* **305**: 422–424.

Wright R. F., E. Lotse, A. Semb (1988a) Reversibility of acidification shown by whole catchment experiments. *Nature* **334**: 670–675.

Wright R. F., S. A. Norton, D. F. Brakke, T. Frogner (1988b) Experimental verification of episodic acidification of freshwaters by sea salts. *Nature* **334**: 422–424.

Wright R. F. and A. Snekvik (1978) Acid precipitation: chemistry and fish populations in 700 lakes in southernmost Norway. *Verh. Internat. Verein. Limnol.* **20**: 765–775.

Index